Modeling of cryogenic sloshing including heat and mass transfer

MODELING OF CRYOGENIC SLOSHING INCLUDING HEAT AND MASS TRANSFER

Vom Fachbereich Produktionstechnik

der

UNIVERSITÄT BREMEN

zur Erlangung des Grades
Doktor-Ingenieur
genehmigte

Dissertation

von

Arnold van Foreest, M.Sc.

Gutachter:

Prof. Dr.-Ing. M. Dreyer
Prof. Dr. A.E.P. Veldman, Universität Groningen, die Niederlande

Tag der mündlichen Prüfung:

06.01.2014

Bibliografische Information der Deutschen Nationalbibliothek

Die Deutsche Nationalbibliothek verzeichnet diese Publikation in der Deutschen Nationalbibliografie; detaillierte bibliografische Daten sind im Internet über http://dnb.d-nb.de abrufbar.

1. Aufl. - Göttingen : Cuvillier, 2014

   Zugl.: Bremen, Univ., Diss., 2014

   978-3-95404-653-9

Zusammenfassung in deutscher Sprache

Ziel der Promotionsarbeit war es einen praktischen Ingenieursansatz aufzustellen mit dem Zweck den Einfluss des Schwappens einer kryogenen Flüssigkeit auf die Wärme- und Stoffübertragung zu untersuchen und die damit verbundenen Temperatur- und Druckänderungen in ihren wesentlichen Zügen mathematisch zu beschreiben. Angewandt auf Trägerraketen lässt sich mithilfe dieses Modells die Druckänderung im Tank vorher sagen, was für eine funktionstüchtige Treibstoffförderung essentiell ist.

Am ZARM wurden Experimente durchgeführt um die Entwicklung des mathematischen Modells zu ermöglichen.

Diese Experimente sind in einem zylindrischen Behälter aus Glas durchgeführt worden. Der Radius dieses Behälters betrug 0.145 m, der Boden hatte die Form einer Halbkugel. Beim Arbeitsfluid handelte es sich um flüssigen Stickstoff, der als Surrogat für die eigentlichen Raketentreibstoffe, flüssiger Wasserstoff und flüssiger Sauerstoff, diente.
Der Behälter wurde geschlossen und bedruckt bis unterschiedliche, jedoch zuvor spezifizierte, Druckniveaus erreicht wurden. Die Bedruckung fand durch zwei unterschiedliche Methoden statt: die aktive Bedruckung und die Selbstbedruckung. Bei der aktiven Bedruckung wurde gasförmiger Stickstoff zugeführt, wohingegen bei der Selbstbedruckung eine Druckerhöhung durch von außen in den Tank eindringender Wärme stattfand. Bei letzterer Methode geht ein Teil des flüssigen Stickstoffs in die Gasphase über und erhöht somit den Tankdruck.
Der Selbstbedrückungsprozess beansprucht mehr Zeit (bis zu einer Stunde) als der aktive Bedrückungsprozess (wenige Minuten).
Nachdem der Druck das gewünschte Niveau erreicht hatte, wurde ein Schwappen der Flüssigkeit durch eine laterale periodische Anregung des Behälters initiiert. Die Anforderungen am Anregungsvorgang waren wie folgt; es musste eine gut definierte Schwappbewegung resultieren und keine chaotischen Wellenformen.
Nach dem Auslösen der Schwappbewegung konnte eine klare Abnahme des Druckes im Behälter beobachtet werden. Der Druck sank bis ein Minimaldruck erreicht war, danach nahm der Druck wieder langsam zu.

Ferner wiesen die Experimente einen starken thermalen Gradienten an der Flüssigkeitsoberfläche nach der Bedruckung auf. Dieser Gradient nahm während des

Schwappens ab und die Temperatur der Flüssigkeitsoberfläche ebenfalls. Eine Analyse der experimentalen Ergebnisse zeigte, dass die Abnahme des Druckes vor allem durch die Kondensation des Gases im Behälter verursacht wurde. Die Kondensation wiederum kann durch die Abnahme der Temperatur an der Flüssigkeitsoberfläche erklärt werden, da dies eine Änderung des Sättigungsdruckes zur Folge hat. Diese experimentellen Befunde weisen klar daraufhin, dass es von Belang ist die Entstehung des Temperaturgradienten zu verstehen und mathematisch modellieren zu können, um somit auch die Druckänderung während der Schwappbewegung zu verstehen.

Das CFD Programm FLOW 3D wurde verwendet, um die Bedrückungsphase zu simulieren und so die Entstehung der Thermalschichtung besser zu verstehen. Es wurde klar, dass es mit FLOW 3D nicht möglich war Ergebnisse zu bekommen, die quantitativ richtig sind, weil die benötigte räumliche Auflösung des numerischen Gitters zu hoch war und deswegen die Rechenzeit eine unakzeptable Dauer von etlichen Monaten erreichte. Die numerischen Ergebnisse waren jedoch ausreichend, um den Wärmetransport innerhalb des Behälter qualitativ zu untersuchen und ein Verständnis für die Phänomene bei der Entstehung der thermischen Schichtung zu schaffen.

Die FLOW 3D Ergebnisse zeigen eindeutig, dass im Falle der Selbstbedruckung der thermale Gradient in der Flüssigkeit hauptsächlich durch die in tangentiale Richtung, d.h. durch die Behälterwand, transportierte Wärme erzeugt wird. Ursache dieses Transportes ist die Tatsache, dass die Wand in der Gasregion wärmer ist als die in der Flüssigkeitsregion, wodurch sich ein Wärmefluss in Richtung der Flüssigkeitsregion aufbaut. Diese Wärme wird gerade unter der Flüssigkeitsoberfläche und somit an die Flüssigkeit abgegeben, was einen thermischen Gradient zur Folge hat.

Im Falle der aktive Bedrückung entsteht der Gradient durch die Erhöhung des Druckes und der damit verbundenen Kondensation an der Phasengrenze (zum erhöhten Druck gehört eine erhöhte Sättigungstemperatur, die an der Flüssigkeitsoberfläche nicht erreicht wird). Die Temperatur an der Flüssigkeitsoberfläche erhöht sich durch die Wärme, die durch die Kondensation freigesetzt wird, bis auf die neue Sättigungstemperatur.

Die Dicke des thermischen Gradienten erhöht sich mit der Zeit durch Wärmeleitung. Die FLOW 3D Ergebnisse lassen vermuten, dass ein einfaches eindimensionales Wärmeleitungsmodel ausreichen könnte, um die Entstehung dieses thermischen Gradienten zu beschreiben. Das Modell berücksichtigt die Dicke der Wand und die der Flüssigkeit durch jeweils eine numerische Zelle, womit es streng genommen ein zweidimensionales Modell ist. Da es sich nur um zwei Zellen in eine Richtung und mehreren hundert in eine andere Raumrichtung handelt, kann mit Recht von einem eindimensionalen Modell gesprochen werden. Der Gradient, der mithilfe dieses eindimensionalen Modells vorhergesagt wird, ist in guter Übereinstimmung mit den experimentell gemessenen Ergebnisse.

Nachdem die Entwicklung eines Modells zur Beschreibung der Entstehung des thermischen Gradient erfolgreich war, konnte als nächster Schritt die Modellierung der Druckänderung beim Schwappen untersucht werden. FLOW 3D konnte hier keinen Beitrag leisten, da es gezeigt wurde das FLOW 3D nicht in der Lage war dieser Prozess zu modellieren wegen einer Beschränkung der Numerik. Aber eine eindimensionale energetische Analyse des Systems zeigte, dass das entwickelte Modell für die Beschreibung der thermischen Gradient auch für die Beschreibung der Druckänderung während Schwappen genutzt werden kann. Hierzu musste der Wärmeleitungskoeffizient in den Regionen erhöht werden, die durch die Schwappbewegung beeinflusst wurden. Diese Erhöhung der Wärmeleitung lässt sich durch die Konvektion, welche durch die Schwappbewegung verursacht wird, erklären. Die Erhöhung des Wärmeleitungskoeffizient wird durch die Nusseltzahl beschrieben, welche ihrerseits aus den experimentellen Ergebnissen gewonnen wird. Die Implementierung dieser Nusseltzahl in das eindimensionale mathematische Modell liefert

Druckänderungen, die in gute Übereinstimmung sind mit den experimentellen Ergebnissen. Die zeitliche Temperaturentwicklung in der Flüssigkeit während des Schwappens ist ebenfalls in gute Übereinstimmung.

Das eindimensionale Modell wurde angewendet auf die Experimente von Moran et. al. [18] und Lacapere et al. [19].

Danach wurde das Modell auf die neue ESC-B Oberstufe angewendet, welche sich momentan bei Astrium in der Entwicklung befindet.

Die Arbeit schließt mit einem Dimensionsanalyse zur Bestimmung der Nusseltzahl. Hier stellt sich heraus, dass der Nusseltzahl eine Funktion ist vom Prandtlzahl und Galileozahl. Es verbleibt zukünftigen Arbeiten diesen Zusammenhang weiter zu untersuchen.

Abstract

Sloshing of liquid propellants in launch vehicle tanks creates unwanted side-effects. These side-effects can be divided in two groups:

- The dynamical phenomenon of sloshing which cause unwanted forces on the launch vehicle which have to be counteracted by the attitude control system.
- The thermodynamical phenomena present during sloshing of a cryogenic liquid which cause pressure changes in the tank.

The dynamical aspects have been extensively studied, but on the thermodynamical side many questions still remain.

This work thus focuses on the latter. The pressure changes caused by sloshing were experimentally investigated at ZARM in a dewar containing liquid nitrogen (as a substitute for the upper stage propellants liquid oxygen and hydrogen). The results of these experiments are discussed and analysed.

The experiments are modelled in FLOW3D, a commercially available Computational Fluid Dynamics (CFD) program. Using the experimental and numerical results, a one dimensional (1D) engineering model is developed which is able to describe the experimental results. This engineering model is then applied to Europe's future upper stage, the ESC-B.

Acknowledgements

In the first place I would like to thank my principal supervisor, Prof. Dr.-Ing. Michael Dreyer, head of the Multiphase Flow Group at ZARM for giving me the opportunity to do my research using experimental results obtained at ZARM and for his active support. I especially acknowledge the fact that he has supported me so actively even though I was an external PhD student.

As a full time employee at DLR-SART I was lucky to get the opportunity to do my PhD on a subject which was also of interest to SART. This way I was able to use some of my office time for my PhD research. Therefore I would like to thank the head of the SART department, Dr. Martin Sippel, for having supported me in this matter.

I am also very grateful for all the support and patience my wife, Venna van Foreest-Cohen, has given me. She was always my motivator, sending me up into my home office at night to "finally write everything down" and never complaining about those lonely evenings she had when I was working late.

Furthermore I would like to thank my father, Dr. Dirk van Foreest, for reading and correcting my report and for his engagement and motivation.

Thanks also to Dr. Farid Gamgami for reading and correcting my report and for the valuable discussions we had.

Table of contents

List of figures

List of tables

Nomenclature

In some cases it was unavoidable or simply practical to give symbols multiple definitions. From the context or subscript it will be clear which definition applies.

a	=	wave amplitude	[m]
a	=	dimensionless vapour velocity	
A	=	constant value	[K]
B	=	constant value	[K^{-1}]
c	=	specific heat (fluid property)	[J kg^{-1} K^{-1}],
C	=	constant value	
D	=	diameter	[m]
f	=	frequency	[Hz]
F	=	fluid fraction	
g	=	gravitational acceleration	[9.81 m s^{-2}]
h	=	heat transfer coefficient	[W m^{-2} K^{-1}]
h	=	specific enthalpy	[J kg^{-1}]
H	=	enthalpy	[J]
H	=	height	[m]
i	=	specific internal energy	[J kg^{-1}]
I	=	internal energy	[J]
I	=	value describing temperature distribution	
J	=	Bessel function	
KE	=	kinetic energy	[J]
L	=	length	[m]
$\tilde{L}$	=	latent heat	[J kg^{-1}]
m	=	mass	[kg]
$\overline{M}$	=	molecular mass	[kg mol^{-1}]
n	=	number of layers	
p	=	pressure	[kPa]
PE	=	potential energy	[J]
q	=	specific heat	[J m^{-2}]
$\tilde{q}$	=	specific heat	[J kg^{-1}]
Q	=	heat	[J]
r	=	radial position	[m]

R	=	radius of cylinder	[m]
$\tilde{R}$	=	gas constant	[J kg^{-1}K^{-1}]
$\bar{R}$	=	universal gas constant	[8.314 J·mol^{-1}K^{-1}]
S	=	surface area	[m^2]
t	=	time	[s]
T	=	temperature	[K]
T	=	temperature sensor	
u	=	velocity	[m s^{-1}]
U	=	volume specific internal energy	[J m^{-3}]
v	=	specific volume	[m^3 kg^{-1}]
V	=	volume	[m^3]
W	=	work	[J]
W	=	weight	[N]
x	=	x direction	[m]
y	=	y direction	[m]
z	=	z direction	[m]

Non dimensional numbers

Bo	=	Bond number
Ga	=	Galileo number
Gr	=	Grashof number
Nu	=	Nusselt number
Pr	=	Prandtl number
Ra	=	Rayleigh number
Re	=	Reynolds number
St	=	Strouhal number

Vectors

a	=	acceleration	[m s^{-2}]
F	=	Force	[N]
n	=	normal vector	
u	=	velocity	[m s^{-1}]

$\dot{q}$ = specific heat flow $\qquad$ [W m^{-2}]
τ = viscous stress $\qquad$ [Pa]

Greek symbols

α = thermal diffusivity $\qquad$ [m^2 s^{-1}]
β = thermal expansion coefficient $\qquad$ [K^{-1}]
β = $2\gamma\omega$ $\qquad$ [rad s^{-1}]
δ = boundary layer thickness $\qquad$ [m]
ε = root of derivative of Bessel function
η = surface displacement $\qquad$ [m]
Δ = delta
Δ = logarithmic decrement
γ = damping ratio
Γ = correcting factor for the bulk gas motion $\qquad$ [m kg J^{-1}s^{-1}]
λ = conduction coefficient $\qquad$ [W m^{-1} K^{-1}]
ρ = density $\qquad$ [kg m^{-3}]
μ = dynamic viscosity $\qquad$ [kg m^{-1} s^{-1}]
ν = kinematic viscosity $\qquad$ [m^2 s^{-1}]
ξ = factor depending on boundary layer profile
ϕ = velocity potential $\qquad$ [m^2 s^{-1}]
Φ = velocity potential of eigenmode $\qquad$ [m^2 s^{-1}]
Ψ = pendulum deflection angle $\qquad$ [rad]
θ = azimuth $\qquad$ [rad]
σ = accommodation coefficient
σ = Stefan-Boltzmann constant $\qquad$ [5.67E-8 W m^{-2} K^{-4}]
ω = frequency $\qquad$ [rad s^{-1}]

Sub/superscripts

0 = in case of ideal gas
1 = fluid #1
2 = fluid #2
a = excitation
A = cell A

b	=	beginning of pressurisation
B	=	cell B
av	=	average value
c	=	characteristic
C	=	cell C
$cond$	=	conduction
$conv$	=	convection
cv	=	control volume
e	=	effective
e	=	exiting
$evap$	=	evaporation
f	=	fill
fr	=	fractional
i	=	i^{th} value
i	=	ingoing
i	=	inner
$infl$	=	reqion of influence
l	=	liquid
$l\text{-}n$	=	entering the liquid passing in normal direction through the dewar wall
ls	=	liquid surface
m	=	mode
max	=	maximum
n	=	wave number
new	=	new value
o	=	outer
old	=	old value
p	=	at constant pressure
p	=	pressurization
pd	=	at maximum rate of pressure drop
pch	=	phase change
$pmin$	=	minimum pressure
s	=	surface
s	=	slosh initiation
sat	=	saturation
ss	=	steady state
T	=	thermal

T = at temperature T

tot = total

u = ullage

v = vapor

v = at constant volume

vap = vaporized

w = wall

x = x direction

y = y direction

z = z direction

Γ = liquid vapour interface

Abbreviations

ap = active pressurization

1D = one dimensional

2D = two dimensional

3D = three dimensional

CFD = Computational Fluid Dynamics

DLR = Deutsches Zentrum für Luft- und Raumfahrt (German Aerospace Center)

EAP = Etage d'Accélération à Poudre

EPC = Etage Principal Cryotechnique

ESC = Etage Supérieur Cryotechnique

GN_2 = gaseous nitrogen

LH_2 = liquid hydrogen

LN_2 = liquid nitrogen

LOX = liquid oxygen

NIST = National Institute of Standards and Technology

SART = Space Launcher Systems Analysis (Systemanalyse Raumtransport)

sp = self-pressurization

VEB = Vehicle Equipment Bay

ZARM = Zentrum für angewandte Raumfahrttechnologie und Mikrogravitation (Center of Applied Spacetechnology and Microgravity)

1   Introduction

Propellant sloshing in launch vehicle tanks has been an important field of research in astronautics since the beginning of the space age. The reason is quite simple. Liquid propellants make up a significant amount of the total mass of a launch vehicle. In launchers where only liquid propellant is used (for example the Delta 4, Saturn V or Falcon launch vehicles from the USA or the Russian Soyuz launcher) liquids account for about 90% of the total mass.

Also in other launchers such as the European Ariane 5 liquid propellant has a significant mass contribution. The Ariane 5 consists of two big solid propellant boosters (EAP), which make up 72% of the total lift off mass. Because the boosters contain solid propellant they are not subjected to sloshing. The main stage (EPC) and the upper stage (ESC-A) contain the liquid propellants LH2 and LOX. The mass of the liquid propellants amounts to 184.9 tons. This accounts for 24% of the total mass at takeoff, which is still a significant amount. A mass breakdown of the Ariane 5 is provided in Table 1-1. A schematic picture of Europe's Ariane 5 launcher is depicted in Figure 1-1 (see also [1]). This picture clearly shows that most of the volume of the launcher is occupied by the liquid propellant tanks.

One can imagine that sloshing can cause significant forces on the vehicle which must be counteracted by the attitude control system. It is therefore very important to understand the sloshing phenomenon and to be able to predict the forces and frequencies introduced by sloshing liquids.

But research of liquid sloshing in containers did not begin at the start of the space age. There are many other areas were liquid motion can be of importance. For example ships transporting containers containing liquids can be subjected to forces exerted by the sloshing liquid. Also the transport of liquids by trucks is an example were sloshing motion might be of importance. However, the influence of the sloshing motion of these two examples can be limited by filling the containers as much as possible. Another example is found in the field of aeronautics. One of the earliest examples is when the wings are filled with propellant and the propellant slosh is coupled with the wing vibration modes, modifying the flutter characteristics [2]. On a larger scale, liquid motion can be of importance in large basins or harbours subjected to tidal influences [3] or earthquakes [4].

Although the above mentioned examples show that liquid slosh is of importance in many other fields, sloshing research got a boost at the start of the space age. Early research on space related sloshing dynamics has been carried out by the USA as well as in the former

Soviet Union. Arguably the best known work on sloshing dynamics is that of Abramson 1966 [5]. This work analyses different sloshing modes for containers of various geometries, as well as damping of the sloshing motion and nonlinear sloshing. Dodge 2000 [6] has revisited the work of Abramson, extending it with more accurate numerical data and knowledge obtained since the 1960s. Other well-known works are those of Miles 1984 [7], developing a weakly nonlinear theory which allows for determination of a phase diagram of sloshing motion, and Ibrahim 2005 [13], summarizing theory and fundamentals of sloshing and containing thousands of references.

	Propellant mass [tons]	Structural mass [tons]
2 x EAP	480	76
EPC	170	13
ESC-A	14.9	2.1
Additional structure*		21
total	664.9	112
Total lift off mass [tons]	777	

*contains all additional structure including payload, interstage, VEB etc.

Table 1-1. Ariane 5 ECA mass breakdown.

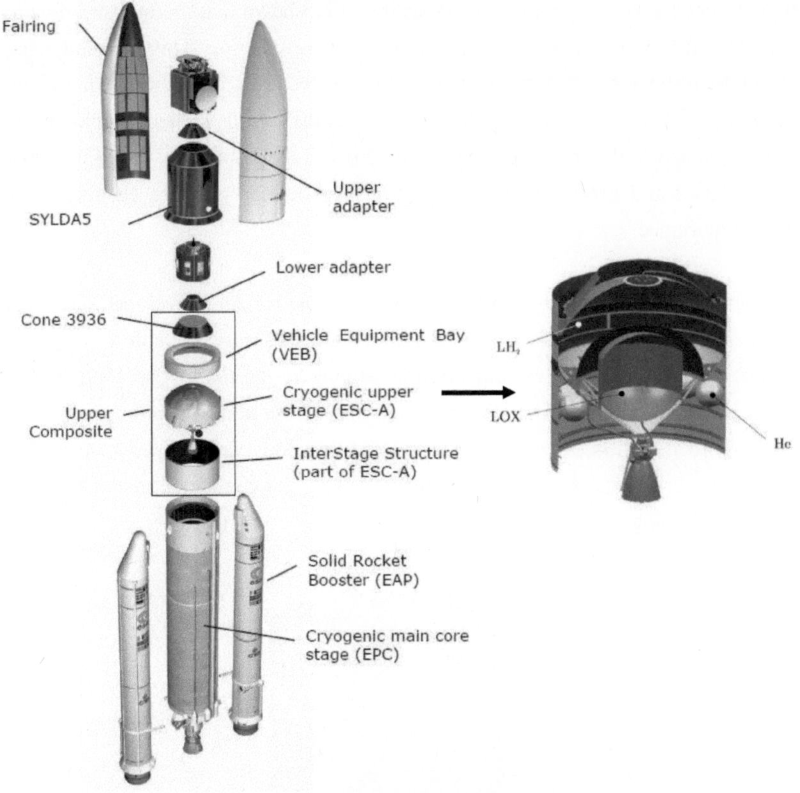

Figure 1-1. Ariane 5 ECA with ESC-A upper stage.

The dynamical behaviour of a sloshing liquid has been extensively studied. However, sloshing can lead to thermodynamic issues as well. This is especially the case when cryogenic propellants are used, such as LOX and LH2. In this case, sloshing is known to cause pressure changes in the tanks. Such pressure changes can be dangerous because launcher tanks are only structurally stable within a certain pressure range. It is therefore important to be able to predict these pressure changes. Pressure changes also have an influence on the required pressurisation gas. The ability to predict the pressure changes would also lead to a more accurate prediction of the required pressurisation gas mass.

Because of the low temperatures of the cryogenic liquid propellants, unavoidable heat leaks from the surroundings enter the propellant and propellant temperature increases. The propellant does not heat up homogeneously but instead thermal layers will form. This is referred to as *thermal stratification*. In this report it will be shown that thermal stratification is the main cause for the pressure changes observed during sloshing. This thermodynamic aspect has not been the subject of many studies. In fact, there are only three experimental campaigns which have resulted in the publication of results on this subject (excluding the experiments conducted in the framework of this PhD research). Especially in missions with long ballistic phases and greater bulk fluid motions, associated pressure changes in the system are of importance.

Currently, a new upper stage for the Ariane 5 launch vehicle is being developed in Europe. This new upper stage, which is called ESC-B (see Figure 1-2), must be able to fly longer missions with long ballistic phases and multiple engine re-ignitions. The long ballistic phases cause considerable propellant heating and the re-ignitions introduce significant sloshing. Therefore it is necessary to understand the thermodynamic aspects of sloshing better. This is why cryogenic sloshing experiments have been conducted at the Centre of Applied Space Technology and Microgravity (ZARM).

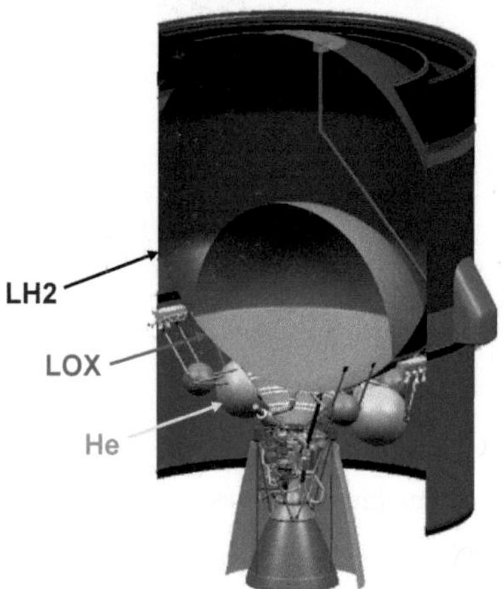

Figure 1-2. ESC-B upper stage.

1.1   State of the art

Analysis of thermal stratification in cryogenic propellants dates back to the beginning of the space age. Much literature is available on this subject, but the work of Chin et al. 1964 [21] is arguably the most commonly referenced. Ring et al. 1964 [8] present an analytical model taking into account liquid outflow. Clark 1965 [9] provides a number of references of work done on stratification available at the time. Some examples of more recent numerical studies into stratification are by Grayson et al. 2006 [10], Schallhorn et al. 2006 [11] and Oliveira et al. 2007 [12].

The influence of sloshing on tank pressure has not been the subject of many studies and is therefore poorly understood. Moran et al. 1994 [18] conducted sloshing experiments using LH2 in a spherical tank with a radius of 0.75 m. Sloshing was initiated by introducing a lateral motion to the tank. Two different excitation modes were selected, one with an amplitude of 1.5 inches (0.0381 m) and a frequency of 0.74 Hz to create chaotic sloshing, the other with an amplitude of 0.5 inches (0.0127 m) and a frequency of 0.95 Hz for stable, planar sloshing. Liquid fill height was varied, resulting in a change in eigenfrequency and resulting liquid response. Moran et al. measured pressure collapse to about 50% of the initial pressure. Results showed a clear dependency on the slosh mode (stable or unstable sloshing).

Lacapere et al. 2007 [19] conducted cryogenic sloshing experiments using LOX and LN_2 as fluids. Experiments were executed using a cryostat with a radius of 0.091 m and a height of 0.8 m. The dewar was filled to 55% of its total volume. Excitation amplitude was 0.003 m and excitation frequency was 2.1 Hz, close to the eigenfrequency of the system. This resulted in very chaotic sloshing. Lacapere et al. measured pressure collapse to about 60% of the initial pressure. Das & Hopfinger 2009 [20] conducted experiments using the liquids FC-72 and HFE 7000. Although not of cryogenic nature, the thermodynamic environment in the system was such that a pressure drop could be produced. Pressure collapse up to 57% of the initial pressure was measured. Das & Hopfinger developed an empirical model which describes pressure drop as a function of an effective diffusion constant and temperature gradient in the liquid. The experimentally determined effective diffusion constant is used as a similarity parameter to scale the experimental results to experimental data obtained by Moran et al. and Lacapere et al., with limited success.

Except for the model of Das & Hopfinger, which only describes maximum pressure drop *rates* but gives no information on total pressure drop in the system, there are no models available which can predict or simulate pressure changes due to sloshing. Much work on the understanding, prediction and simulation of pressure behaviour remains to be done.

1.2   Goal of the research and the structure of this report

The ultimate goal is to develop simple engineering models which can be used to predict the influence of sloshing on pressure. The first step is to improve the understanding of this phenomenon. Therefore sloshing experiments have been conducted using LN2. The experiments were conducted at ZARM in Bremen, Germany. The experiments were modelled using the commercially available software FLOW 3D, to create a better understanding of the physical processes involved. Once the problem is understood, the extraction of simple engineering models can be attempted. The remainder of this report is structured as follows:

- Chapter 2 presents and discusses the basic theoretical background required for this research.
- Chapter 3 describes the experimental setup and experimental results.
- Chapter 4 discusses the numerical analysis of thermal stratification and the development of an engineering model to describe thermal stratification.
- Chapter 5 elaborates on the numerical analysis of the influence of sloshing on the pressure.
- Chapter 6 provides a summary of this report.
- Chapter 7 contains the conclusions and recommendations.

2 Theoretical background & research status

This chapter provides an overview of the theoretical background applicable to the research. The focus of the chapter will lie on thermodynamics and heat transfer because experimental results are mainly explained by thermodynamic processes.

A general overview of the relevant fluid dynamics theory and equations will be given and elaborated by special cases. The special cases presented are of direct interest to this research.

The dynamics of sloshing will be discussed only briefly, as this research is not focussed on investigating dynamical features. However, to better understand the experimental results, it is important to have some basic knowledge of the dynamics. The reader is referred to [5] [6] [13] for more details on the dynamics of sloshing.

2.1 Fundamental fluid dynamics and thermodynamics equations

2.1.1 First law of thermodynamics

The first law of thermodynamics describes the change of the (internal) energy in a system due to heat added to a system and work done on this system. The first law of thermodynamics will be used in this research to determine heat input into the experimental dewar.

Energy balance for closed systems

A closed system is a system where no mass enters or exits the system. The energy balance for a closed system results in:

$$dE = dKE + dPE + dI = dQ - dW \qquad \text{2-1}$$

where the energy E has been subdivided in kinetic energy KE, potential energy PE and internal energy I. Q represents heat entering the system and W the work done on the system. In engineering thermodynamics the sign convention is usually as follows:

- $dQ > 0$ when heat enters the system
- $dQ < 0$ when heat leaves the system
- $dW > 0$ if work is done by the system
- $dW < 0$ if work is done on the system

This sign convention is chosen because in most thermodynamic systems encountered in engineering, the heat entering the system is transferred into work done by the system. In this nominal case values of heat and work are thus positive.

Neglecting kinetic and potential energy effects equation 2-1 can be rewritten as:

$$dQ = dI + dW = dI + pdV \qquad\qquad 2\text{-}2$$

where V is the volume. In mass specific terms this can be written as:

$$d\tilde{q} = di + dw = di + pdv \qquad\qquad 2\text{-}3$$

In appendix IV it is shown that this can also be written as:

$$d\tilde{q} = di + pdv = c_v(T,v)dT + T\left(\frac{\partial p}{\partial T}\right)_v dv \qquad\qquad 2\text{-}4$$

where c_v is the specific heat of a fluid at constant volume. For ideal gases this reduces to:

$$d\tilde{q} = di + pdv = c_v^0(T)dT + pdv \qquad\qquad 2\text{-}5$$

In some cases it is convenient to introduce the enthalpy H:

$$H = I + pV \qquad\qquad 2\text{-}6$$

In mass specific terms this can be written as:

$$h = i + pv \qquad\qquad 2\text{-}7$$

The derivative gives $dH = dI + pdV + Vdp$. Using this, equation 2-2 can be rewritten as

$$dQ = dH - Vdp \qquad\qquad 2\text{-}8$$

Or in mass specific terms:

$$d\tilde{q} = dh - vdp \qquad\qquad 2\text{-}9$$

In appendix IV it is shown that this can also be written as:

$$d\tilde{q} = dh - vdp = c_p(T,p)dT - T\left(\frac{\partial v}{\partial T}\right)_p dp \qquad\qquad 2\text{-}10$$

where c_p is the specific heat of a fluid at constant pressure. For ideal gases this reduces to:

$$d\tilde{q} = c_p^0(T)dT - vdp \qquad\qquad 2\text{-}11$$

c_p and c_v in liquids

For incompressible substances, $c_p = c_v$ and equations 2-4 and 2-10 can be written as $d\tilde{q} = c_v dT = c_p dT$. In literature it is often stated that because liquids are almost incompressible, these two values will be almost equal. This is **wrong**. Although the specific volume v shows only very little variation with temperature or pressure, this does not mean that c_p and c_v are equal. In fact, in the case of liquid nitrogen, the difference between c_p and c_v can be as large as a factor 1.9. Using the correct value of specific heat is important for the 1D engineering models (chapters 4 and 5). A discussion on the difference between the two specific heats can be found in appendix IV. The reader is urged to study this appendix carefully.

Energy balance for open, unsteady systems

In open systems mass enters and/or exits the system. Using a control volume for analysing the system under consideration the energy balance for a control volume results in:

$$\dot{E}_{cv} = \dot{KE}_{cv} + \dot{PE}_{cv} + \dot{I}_{cv} = \dot{Q} - \dot{W} + \sum_i \dot{m}_i\left(i_i + \frac{u_i^2}{2} + gz_i\right) - \sum_e \dot{m}_e\left(i_e + \frac{u_e^2}{2} + gz_e\right) \qquad 2\text{-}12$$

where the terms $\dot{m}\dfrac{u^2}{2}$ and $\dot{m}gz$ represent the changes to the kinetic and potential energy by the incoming and/or exiting flow. The work W in an open system consists of two parts. The first part is the work introduced at the inlets and exits where mass enters or exits the control volume and the second part is all the other work on the control volume (also referred to as technical work). This means the work can be written as $\dot{W} = p_e S_e u_e - p_i S_i u_i + \dot{W}_{cv}$, where the first two terms involve the work done by the entering and exiting mass and u_e and u_i are the exit and inlet velocities respectively. Because $Su = \dot{m}v$ (where v is the specific volume) and the specific enthalpy h is defined by $h = i + pv$, the above equation can be rewritten in terms of specific enthalpy of the entering and exiting flow:

$$\dot{E}_{cv} = \dot{KE}_{cv} + \dot{PE}_{cv} + \dot{I}_{cv} = \dot{Q}_{cv} - \dot{W}_{cv} + \sum_i \dot{m}_i \left(h_i + \frac{u_i^2}{2} + gz_i \right) - \sum_e \dot{m}_e \left(h_e + \frac{u_e^2}{2} + gz_e \right) \qquad 2\text{-}13$$

By neglecting potential and kinetic energy effects the following relation is obtained:

$$\dot{I}_{cv} = \dot{Q}_{cv} - \dot{W}_{cv} + \sum_i \dot{m}_i h_i - \sum_e \dot{m}_e h_e \qquad\qquad 2\text{-}14$$

2.1.2　Fundamental fluid dynamics equations

To understand convective heat transport and phase change processes better requires an understanding of the fundamental fluid dynamics equations. Three fundamental equations exist, namely the conservation of mass equation, the conservation of momentum equation and the conservation of energy equation. These are presented below under the following assumptions:

1. Single species system
2. Viscosity, heat conductivity and heat capacity are constant
3. In the energy equation kinetic effects can be neglected as well as viscous heating and heating due to body forces
4. Newtonian behaviour of fluid
5. No slip at the liquid vapour interface
6. No slip at walls
7. No temperature jumps
8. No interface resistance for phase changes
9. Temperature at free surface is equal to the saturation temperature

For each equation the applicable boundary conditions will be provided. This includes boundary conditions at a liquid vapour interface.

2.1.2.1 Conservation of mass

The conservation of mass requires:

$$\frac{\partial \rho}{\partial t} + \nabla \cdot (\rho \mathbf{u}) = \frac{\partial \rho}{\partial t} + \mathbf{u} \cdot \nabla \rho + \rho \nabla \cdot \mathbf{u} = 0 \qquad \text{2-15}$$

For incompressible *flow* the density is constant within an infinitesimal volume that moves with the velocity of the fluid. Thus the density field does not have to be uniform. It is a property of the flow and not the fluid. Incompressible flow is achieved when $\nabla \cdot \mathbf{u} = 0$ yielding:

$$\frac{\partial \rho}{\partial t} + \mathbf{u} \cdot \nabla \rho = 0 \qquad \text{2-16}$$

For an incompressible *fluid*, ρ is a fixed constant and is a property of the fluid (the density field is uniform and does not change in time) which implies:

$$\frac{\partial \rho}{\partial t} = 0$$
$$\nabla \rho = 0 \qquad \text{2-17}$$

By substitution in equation 2-15, $\nabla \cdot \mathbf{u} = 0$ is also obtained for incompressible fluids. Therefore incompressible fluids always undergo flow that is incompressible, but the converse is not true.

In literature it is often stated that for incompressible flow the density is constant. Even though this is technically incorrect, this statement seems general practice.

Boundary conditions

At the liquid vapour interface the mass conservation requires:

$$\rho_l (\mathbf{u}_l - \mathbf{u}_\Gamma) \cdot \mathbf{n}_{ls} = \rho_v (\mathbf{u}_v - \mathbf{u}_\Gamma) \cdot \mathbf{n}_{ls} = \frac{\dot{m}_{pch}}{S_{ls}}$$

The normal vector $\mathbf{n}_{ls}$ at the liquid surface is positive pointing out of the liquid. Here, the phase change mass flux has been approximated by the total phase change mass flow rate divided through the liquid vapour interface area.

2.1.2.2 Conservation of momentum

The conservation of momentum for a system undergoing an external acceleration $\mathbf{a}$ requires:

$$\rho(\frac{\partial \mathbf{u}}{\partial t} + \mathbf{u} \cdot \nabla \mathbf{u}) = -\nabla p + \nabla \cdot \boldsymbol{\tau} + \rho \mathbf{a} - \rho \beta (T - T_\infty)\mathbf{a} \qquad \text{2-18}$$

where $\boldsymbol{\tau}$ is the stress tensor and the last term on the right hand side describes buoyancy, an effect which is elaborated on in section 2.1.3.2. For incompressible flow this equation reduces to:

$$\rho(\frac{\partial \mathbf{u}}{\partial t} + \mathbf{u} \cdot \nabla \mathbf{u}) = -\nabla p + \mu \nabla^2 \mathbf{u} + \rho \mathbf{a} - \rho \beta (T - T_\infty)\mathbf{a} \qquad \text{2-19}$$

Boundary conditions

At the liquid vapour interface the momentum conservation requires:

$$\rho_l\left((\mathbf{u}_l - \mathbf{u}_\Gamma) \cdot \mathbf{n}_{ls}\right)(\mathbf{u}_l \cdot \mathbf{n}_{ls} - \mathbf{u}_v \cdot \mathbf{n}_{ls})\mathbf{n}_{ls} + p_v - p_l - \mathbf{n}_{ls} \cdot (\boldsymbol{\tau}_v - \boldsymbol{\tau}_l) \cdot \mathbf{n}_{ls} = \sigma(\nabla_S \cdot \mathbf{n}_{ls})\mathbf{n}_{ls} - \sigma_T(\nabla_S T_l)$$

where σ is the surface tension coefficient and ∇_S is the surface divergence.

At walls obviously the boundary condition $\mathbf{u} \cdot \mathbf{n}_w = 0$ is obtained. Furthermore, because of the no slip assumption $\mathbf{u} \times \mathbf{n}_w = 0$. Together this means that at walls $\mathbf{u} = 0$ is obtained. The normal vector $\mathbf{n}_w$ at the wall is positive pointing into the liquid.

2.1.2.3 Conservation of energy

The conservation of energy requires:

$$\frac{\partial}{\partial t}\rho i + \nabla \cdot (\rho i \mathbf{u}) = -\nabla \cdot (p\mathbf{u}) + \nabla \cdot (\lambda \nabla T)$$

which can also be written as:

$$i\frac{\partial \rho}{\partial t} + \rho\frac{\partial i}{\partial t} + \mathbf{u} \cdot \nabla(\rho i) + \rho i \nabla \cdot \mathbf{u} = -\nabla \cdot (p\mathbf{u}) + \nabla \cdot (\lambda \nabla T)$$

By using equation 2-15 (conservation of mass) the energy equation reduces to:

$$\rho \frac{\partial i}{\partial t} = -\mathbf{u} \cdot \rho \nabla i - \nabla \cdot (p\mathbf{u}) + \nabla \cdot (\lambda \nabla T) \qquad \text{2-20}$$

The first and second term on the right hand side can be written as:

$$\mathbf{u} \cdot \rho \nabla i + \nabla \cdot (p\mathbf{u}) = \mathbf{u} \cdot \rho \nabla i + \mathbf{u} \cdot \nabla p + p \nabla \cdot \mathbf{u} .$$

For incompressible fluids this reduces to:

$$\mathbf{u} \cdot \rho \nabla i + \mathbf{u} \cdot \nabla p$$

If $\mathbf{u} \cdot \nabla p$ can be assumed small this equation results in (see equation IV-9):

$$\mathbf{u} \cdot \rho \nabla i + p \nabla \cdot \mathbf{u} = \rho c_p \mathbf{u} \cdot \nabla T$$

For liquids the term on the left hand side of 2-20 can be modelled as $\rho \dfrac{\partial i}{\partial t} = \rho c_p \dfrac{\partial T}{\partial t}$ (see equation IV-9). Thus, for incompressible fluids and liquids equation 2-20 then reduces to:

$$\rho c_p \left(\frac{\partial T}{\partial t} + \mathbf{u} \cdot \nabla T \right) = \lambda \nabla^2 T \qquad \text{2-21}$$

Boundary conditions

First of all assumption 7 requires that $T_l = T_v$. Furthermore assumption 9 requires that $T_l = T_v = T_{sat}(p_v)$.

At the liquid vapour interface the energy conservation requires:

$$\tilde{L} \rho_l (\mathbf{u}_l - \mathbf{u}_\Gamma) \cdot \mathbf{n}_{ls} = \tilde{L} \frac{\dot{m}_{pch}}{S} = \lambda_v \frac{\partial T_v}{\partial \mathbf{n}_{ls}} - \lambda_l \frac{\partial T_l}{\partial \mathbf{n}_{ls}} \qquad \text{2-22}$$

At walls it requires $\dot{\mathbf{q}} = -\lambda \dfrac{\partial T}{\partial \mathbf{n}_w}$. Sometimes this is determined as

$\dot{\mathbf{q}} = -\lambda \dfrac{\partial T}{\partial \mathbf{n}_w} = h(T_w - T_\infty)$, where h is the heat transfer coefficient and T_∞ is the bulk fluid temperature.

In this research the energy equation and heat transfer will be of particular importance. Therefore a more detailed elaboration on heat transfer is given in section 2.2.1.

2.1.3 Special cases of the fluid dynamics equations

2.1.3.1 Boundary layer equations

Because of the no slip condition, the fluid near the wall will have a smaller velocity than the bulk fluid further away from the wall. The layer of reduced velocity is the velocity boundary layer. A common definition is that the thickness of the boundary layer is the thickness of the fluid layer between the surface and that point in the fluid where $u = 0.99u_\infty$ (99% of the bulk velocity).

Fluid flows in boundary layers are governed by simplified forms of the fundamental equations, when the so called boundary layer approximations are valid. For simplicity only a two dimensional flow is considered which is very thin in the y direction ($\delta \!<\!\!< \! L_c$). The boundary layer approximations are that the flow occurs primarily in the x direction and $u_y \ll u_x$, and changes in properties in the downstream direction are much smaller compared to the cross stream direction and pressure is constant in the cross stream direction. A schematic view of a boundary layer is given in Figure 2-1.

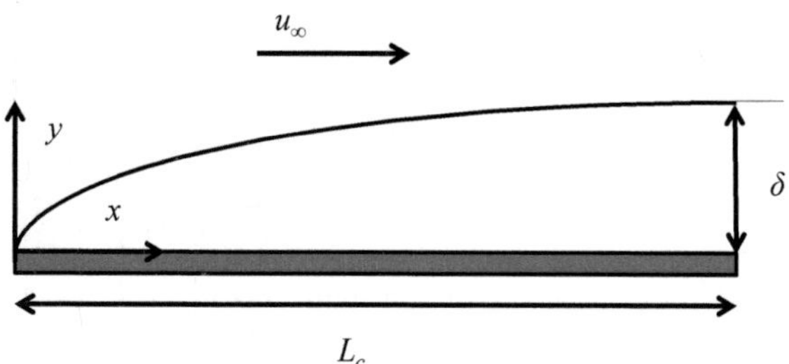

Figure 2-1. Boundary layer over a flat plate.

The equations can be made dimensionless (indicated by *) by dividing through the relevant quantities:

$$x^* = \frac{x}{L_c}, \; y^* = \frac{y}{L_c}, \; u_x^* = \frac{u_x}{u_\infty}, \; u_y^* = \frac{u_y}{u_\infty}, \; p^* = \frac{p}{\rho u_\infty^2}, \; T^* = \frac{T - T_s}{T_\infty - T_s}, \; t^* = \frac{t}{L/u_\infty} \qquad \text{2-23}$$

where the characteristic length L_c is the plate length, the characteristic velocity u_∞ is the bulk fluid velocity, the characteristic pressure is the bulk dynamic pressure and the characteristic temperature difference is the difference between bulk temperature and surface temperature. By applying the boundary layer approximations, assuming incompressible fluid and subsequentually making the equations dimensionless the following results are obtained:

$$\frac{\partial u_x}{\partial x} + \frac{\partial u_y}{\partial y} = 0 \quad \text{(conservation of mass equation)}$$

$$\frac{\partial u_x^*}{\partial x^*} + \frac{\partial u_y^*}{\partial y^*} = 0 \quad \text{(dimensionless conservation of mass equation)}$$

2-24

$$\rho\left(\frac{\partial u_x}{\partial t} + u_x\frac{\partial u_x}{\partial x} + u_y\frac{\partial u_x}{\partial y}\right) = \mu\frac{\partial^2 u_x}{\partial y^2} - \frac{\partial p}{\partial x} \quad \text{(conservation of } x \text{ momentum}$$

equation)

$$\left(\frac{\partial u_x^*}{\partial t^*} + u_x^*\frac{\partial u_x^*}{\partial x^*} + u_y^*\frac{\partial u_x^*}{\partial y^*}\right) = \frac{1}{\mathrm{Re}_{Lc}}\frac{\partial^2 u_x^*}{\partial y^{*2}} - \frac{\partial p^*}{\partial x^*} \quad \text{(dimensionless}$$

2-25

conservation of x momentum equation)

$$\rho c_p\left(\frac{\partial T}{\partial t} + u_x\frac{\partial T}{\partial x} + u_y\frac{\partial T}{\partial y}\right) = \lambda\left(\frac{\partial^2 T}{\partial y^2}\right) \quad \text{(conservation of energy equation)}$$

$$\frac{\partial T^*}{\partial t^*} + u_x^*\frac{\partial T^*}{\partial x^*} + u_y^*\frac{\partial T^*}{\partial y^*} = \frac{1}{\mathrm{Re}_{Lc}\,\mathrm{Pr}}\left(\frac{\partial^2 T^*}{\partial y^{*2}}\right) \quad \text{(dimensionless conservation of}$$

2-26

energy equation)

From the equations in dimensionless form it is seen that Reynolds number and Prandtl number are the relevant similarity parameters.

2.1.3.2   Buoyancy

In systems experiencing acceleration, a low density body placed in a higher density fluid will experience a force in the opposite direction of the accelerative force. This effect is called buoyancy. The net force of a body experiencing a buoyancy force is given by

$$\mathbf{F}_{net} = \mathbf{W} - \mathbf{F}_{buoyancy} = \rho_{body}V_{body}\mathbf{a} - \rho_{fluid}V_{body}\mathbf{a} = \left(\rho_{body} - \rho_{fluid}\right)V_{body}\mathbf{a}$$

where V_{body} is the volume of the body. If a certain region of a fluid is heated the fluid will expand and density will decrease. This process, called *thermal expansion* can be described by the following relation:

$$\beta = \frac{1}{v}\left(\frac{\partial v}{\partial T}\right)_p = -\frac{1}{\rho}\left(\frac{\partial \rho}{\partial T}\right)_p \qquad\qquad\qquad 2\text{-}27$$

where β is the thermal expansion coefficient and v is the specific volume. The net force can thus be expressed in differential form as $\rho\beta(T - T_\infty)\mathbf{a}$.

A typical case consists of a fluid in contact with a hot vertical wall. The fluid will start to heat up and expand. If there is gravity acting in the vertical direction a buoyancy force will act on the fluid and the fluid will start to rise upward. The boundary layer momentum equation (equation 2-25) can be adapted for this specific case, assuming gravity acts in the negative x direction (coordinates according to Figure 2-1):

$$\rho\left(u_x \frac{\partial u_x}{\partial x} + u_y \frac{\partial u_x}{\partial y}\right) = \mu \frac{\partial u_x^2}{\partial y^2} + (\rho_\infty - \rho)g = \mu \frac{\partial u^2}{\partial y^2} + \rho g\beta(T_s - T_\infty)$$

Or in dimensionless form: $\qquad\qquad\qquad\qquad\qquad\qquad\qquad\qquad 2\text{-}28$

$$u_x^* \frac{\partial u_x^*}{\partial x^*} + u_y^* \frac{\partial u_x^*}{\partial y^*} = \left(\frac{g\beta(T_s - T_\infty)L_c^3}{v^2}\right)\frac{T^*}{\mathrm{Re}_{L_c}^2} + \frac{1}{\mathrm{Re}_{Lc}}\frac{\partial u_x^{*2}}{\partial y^{*2}}$$

The term $\dfrac{g\beta(T_s - T_\infty)L_c^3}{v^2}$ is called the Grashof number and is the relevant similarity parameter in natural convection driven flows ($\mathrm{Gr}_{L_c} \gg \mathrm{Re}_{L_c}^2$):

$$\mathrm{Gr}_{L_c} = \frac{g\beta(T_s - T_\infty)L_c^3}{v^2} \qquad\qquad\qquad 2\text{-}29$$

Sometimes it is more convenient to define a modified Grashof number Gr^*_{Lc}:

$$\mathrm{Gr}_{Lc}^* = \frac{g\beta\,\dot{q}\,L_c^4}{\lambda v^2} \qquad\qquad\qquad 2\text{-}30$$

Where $Gr^*_{Lc} = NuGr_{Lc}$ and $\dot{q}$ is the area specific heat flux.

2.2   Heat transfer: a boundary condition for the energy equation

2.2.1   Heat transfer

Heat transfer to a material can be realized by three well known mechanisms:

Radiation

This type of heat transfer is caused by electromagnetic waves, radiated by heated bodies. It does not require the presence of an intervening medium. The heat radiated by a body at a certain temperature is described by:

$$\dot{q}_{rad} = \sigma T_s^4 \qquad\qquad 2\text{-}31$$

where σ is the Stefan-Boltzmann constant (5.67E-8 W m^{-2} K^{-4}).

Conduction

Heat is conducted through a material when a temperature gradient is present. The heat transfer rate is given by:

$$\dot{\mathbf{q}}_{cond} = -\lambda \nabla T \qquad\qquad 2\text{-}32$$

where λ is the heat conduction coefficient.
Heat conductivity can also be expressed by using the diffusion constant α:

$$\alpha = \frac{\lambda}{\rho c_p} \qquad\qquad 2\text{-}33$$

Convection

This describes the heat transfer between a solid surface and the adjacent liquid or gas which is in motion. It thus involves the combined effects of conduction and fluid motion. Fluid motion increases the conductive heat transfer because the influences the temperature gradient, but it also complicates the determination of the heat transfer rate. Convection is called *forced convection* if the fluid motion is introduced by external means, for example a pump, a fan or wind. Convection is called natural convection if fluid motion is introduced by buoyancy effects caused by density differences in the fluid due to variations in the temperature. Convective heat transfer is expressed as:

$$\dot{q}_{conv} = h(T_s - T_\infty)$$

2-34

where h is the experimentally determined convection heat transfer coefficient (it is not a fluid property), T_s is the temperature of the solid surface and T_∞ is the temperature of the fluid sufficiently far away from the surface. Note that this equation can also be used for cases where the fluid is stationary but there is no information on the temperature gradient appearing in 2-32 .

Because it will be shown that thermal stratification in cryogenic liquids is of major importance to describe pressure response during sloshing, a closer look is taken into natural convection.

The no slip condition and the no temperature jump condition result in the fact that heat transfer between solid and fluid at the solid surface takes place by pure conduction. Using the 2 D case in Figure 2-1 as a reference this results in:

$$\dot{q}_{conv} = \dot{q}_{cond} = -\lambda \frac{dT}{dy}\bigg|_{y=0}$$

2-35

Combining this equation with equations 2-32 and 2-34 yields the convection heat transfer coefficient:

$$h = \frac{\lambda \dfrac{dT}{dy}\bigg|_{y=0}}{T_s - T_\infty}$$

2-36

In convective heat transfers studies, the Nusselt number is used as a dimensionless form of the convection heat transfer coefficient. The Nusselt number is defined as follows:

$$\mathrm{Nu} = \frac{hL_c}{\lambda}$$

2-37

where L_c is a characteristic length. The Nusselt number represents the enhancement of the heat transfer through a fluid layer by convection compared to the heat transported by conduction in a stationary fluid over the same fluid layer.

Similarly to the velocity boundary layer discussed in section 2.1.3.1, if a fluid with a certain bulk temperature flows along a surface with a different temperature a thermal boundary will form because of the no temperature jump condition. Also in this case the boundary layer thickness is defined as the thickness of the fluid layer between the surface and that point in the fluid where $T - T_s = 0.99(T_\infty - T_s)$. The relative thickness of the velocity and thermal boundary layers is given by the Prandtl number:

$$\text{Pr} = \frac{\nu}{\alpha} = \frac{\mu c_p}{\lambda} = \text{momentum diffusivity/thermal diffusivity} \qquad 2\text{-}38$$

Gases usually have a Prandtl number in the order of 1, but Prandtl numbers of liquid vary from 0.01 (liquid metals) to 100,000 (heavy oils).

Flows may be of laminar or turbulent nature. The convection heat transfer coefficient of turbulent flow is usually larger than that of laminar flow. Also turbulent flows have a thicker boundary layer. The transfer from laminar to turbulent flow often occurs when a critical Reynolds number has been reached. The Reynolds number is a dimensionless number with relates inertia forces to viscous forces:

$$\text{Re} = \frac{u L_c}{\nu} = \frac{\rho u L_c}{\mu} \qquad 2\text{-}39$$

2.2.2 Heat transfer in boundary layers

In most cases, fluid motion is so complex that no analytical solutions can be obtained to determine the heat transfer between a solid wall and a boundary layer. Heat transfer relations are therefore mostly empirical and built on extensive experimental studies. In case of forced convection boundary layers, simple empirical relations for the average Nusselt number over a certain surface are often given by:

$$\text{Nu}_{Lc} = C\,\text{Re}_{Lc}^{n}\,\text{Pr}^{m} \qquad 2\text{-}40$$

where C is a constant.

A number which is often used in literature for free convection boundary layers is the Rayleigh number, which is defined as the product of Grashof and Prandtl number:

$$\mathrm{Ra}_{Lc} = \mathrm{Gr}_{Lc}\,\mathrm{Pr} \qquad\qquad 2\text{-}41$$

Empirical heat transfer relations for flow along a vertical wall are given in Table 2-1. Nusselt numbers concern the average values over the complete characteristic length Lc, which in this case is the plate length. The results can be rewritten in terms of the modified Grashof number in case the heat flux into the fluid is known but not the temperature difference T_s-T_∞. For example $\mathrm{Nu} = 0.59\mathrm{Ra}_{Lc}^{0.25}$ can be written as $\mathrm{Nu} = 0.59\left(\dfrac{\mathrm{Ra}_{Lc}^{*}}{0.59}\right)^{0.2}$

where $\mathrm{Ra}_{Lc}^{*} = \mathrm{Gr}_{Lc}^{*}\,\mathrm{Pr} = \mathrm{Gr}_{Lc}\,\mathrm{Pr}\,\mathrm{Nu}$.

The Nusselt number depends on the flow conditions. The transition from laminar to turbulent flow occurs around Ra_{Lc} = 1E09 or Ra*$_{Lc}$ = 1E11 [21] in case of a free convective boundary layer. In case of a forced convective boundary layer the transition from laminar to turbulent flow is taken to occur around Re_{Lc} = 5E05 [23]. Note that for free convection the Rayleigh number is the reference and in forced convection it is the Reynolds number.

Free convection	
According to [23]	$Nu = 0.59 Ra_{Lc}^{1/4}$ $10^4 < Ra_{Lc} < 10^9$ $Nu = 0.1 Ra_{Lc}^{1/3}$ $10^9 < Ra_{Lc} < 10^{13}$
According to [24]	$Nu = \left(0.825 + \dfrac{0.387 Ra_{Lc}^{1/6}}{\left(1 + 0.492 Pr^{9/16}\right)^{8/27}} \right)^2$ $10^4 < Ra_{Lc} < 10^{13}$
According to [25]	$Nu = \dfrac{4}{3} \left(\left(\dfrac{Gr_{Lc}}{4} \right)^{1/4} \dfrac{0.849 Pr^{1/2}}{\left(1 + 2.006 Pr^{1/2} + 2.034\right)^{1/4}} \right)$ $Ra_{Lc} < 10^9$
According to [26]	$Nu = 0.13 Ra_{Lc}^{1/3}$ $Ra_{Lc} > 10^9$
Forced convection	
According to [29]	$Nu = 0.664 Re_{Lc}^{1/2} Pr^{1/3}$ $0.6 < Pr < 10$, laminar flow
According to [29]	$Nu = \dfrac{0.037 Re_{Lc}^{0.8} Pr}{1 + 2.443 Re_{Lc}^{-0.1} \left(Pr^{2/3} - 1\right)}$ $0.6 < Pr < 2000$, turbulent flow, $5E05 < Re_{Lc} < 1E07$

Table 2-1. Average Nusselt number relations for convective flow along a vertical wall.

The relations regarding free convective boundary layers listed in the table above are compared to each other in Figure 2-2. The Prandtl number is assumed at Pr =2.2, compliant with LN2. The reference length has been set to L_c = 1m. In the turbulent region the difference between the maximum and minimum predicted values amounts to 30%, where the highest values are predicted by the relation according to [26] and the minimum values are predicted by the relation according to [23].

The relation according to [24] covers the turbulent and laminar range using one equation. This relation is the preferred one for free convective boundary layers.

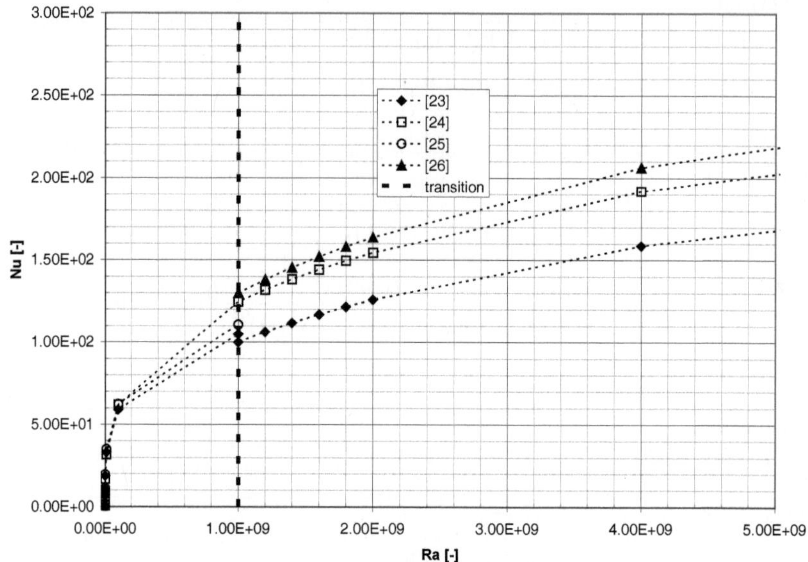

Figure 2-2. Nusselt number relations for a free convection boundary layer with Pr = 2.2 and L_c = 1m.

The relation according to [29] is depicted in Figure 2-3. The Prandtl number is assumed at Pr = 2.2, compliant with LN2. The reference length has been set to L_c = 1m. At the transition Reynolds number a jump in the Nusselt number occurs which clearly shows the impact of a turbulent boundary layer on the heat transfer.

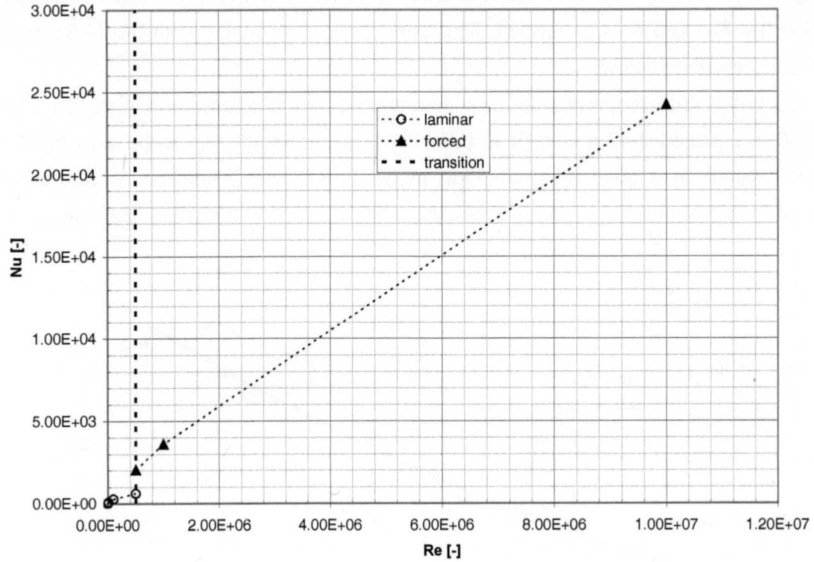

Figure 2-3. Nusselt number relations for a forced convection boundary layer with Pr = 2.2 and L_c = 1m.

2.3 Phase change

2.3.1 Saturation curve

Phase changes can occur when the temperature of a fluid is not equal to the saturation temperature. If a liquid is heated above its saturation temperature, evaporation or boiling can occur. If a gas is cooled below its saturation temperature, condensation can occur. The saturation temperature depends on pressure. The pressure at which a fluid at a certain temperature can undergo a phase change is called the saturation pressure.

The saturation temperature can be calculated by the Clausius-Clapyron equation [37]:

$$T_{sat} = \left[\frac{1}{T_{vl}} - B \ln\left(\frac{p_{sat}}{p_{vl}} \right) \right]^{-1} \qquad 2\text{-}42$$

where T_{v1} and p_{v1} define a point on the saturation curve (for example T_{v1} = 77.24 K and p_{v1} = 100 kPa for N_2) and B is a constant. This constant is determined from the NIST fluid properties as described below.

For N_2 the NIST fluid properties define a saturation curve as shown in Figure 2-4 . By adapting B in equation 2-42 a best fit to this curve can be found. As can be seen in, B = 0.0014 K^{-1} results in a good approximation to the NIST data. Points below this curve correspond to the vapour phase, points above the curve correspond to the liquid phase.

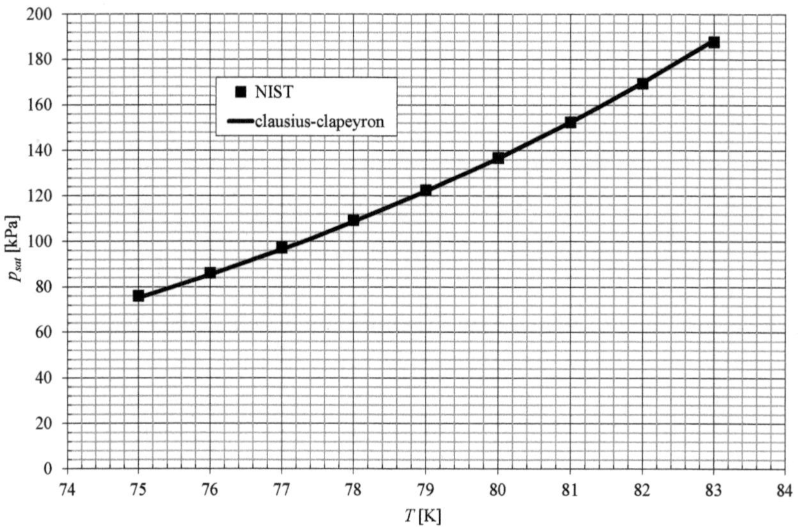

Figure 2-4. Saturation curve for N2 according to NIST and Clausius-Clapeyron with B = 0.0014 K^{-1}.

2.3.2 Evaporation and boiling

In general two types of phase changes can be observed, namely phase changes over a liquid-vapour free surface and phase changes which occur at a solid wall. Phase change over a free surface will occur if the saturation pressure of the liquid is not equal to the saturation pressure of the vapour. For example water in a lake at 20°C will evaporate when air temperature is also 20°C and humidity is 60%, because the liquid saturation pressure is about 2.3 kPa but the vapour saturation pressure is only 1.4 kPa. Phase changes over a liquid-vapour surface where liquid transforms into vapour is called evaporation.

Boiling occurs at a solid wall. This happens when the wall temperature is higher than the saturation temperature of the total pressure. For water at 1 bar, boiling will start if the wall temperature is higher than 100°C.

2.3.3　Condensation

One of the most common examples provided in literature on the topic of condensation is Nusselt's thin film theory [23][27][29]. This theory describes the condensation of vapour at a vertical, subcooled wall. If a liquid film of condensed vapour wets a cold wall, the vapour will condensate at the liquid surface. The latent heat of vaporization must be removed from the liquid surface to sustain the process, so the rate of condensation is linked to the heat transfer rate over the liquid film from interface to wall. There are two ways to analyse this problem, one is an integral analysis and the other is a boundary layer analysis.

Integral analysis
When the following assumptions are made:
- Laminar flow
- Constant properties
- Subcooling of liquid is negligible in the energy balance
- Inertia effect are negligible in the momentum balance
- The vapour is stationary and exerts no drag
- The liquid-vapour interface is smooth
- Heat transfer across the film is by conduction only (convection is neglected)

then a force balance on a liquid element yields (Figure 2-5):

$$(\delta - y)dx(\rho_l - \rho_v)g = \mu_l\left(\frac{du}{dy}\right)dx \qquad\qquad 2\text{-}43$$

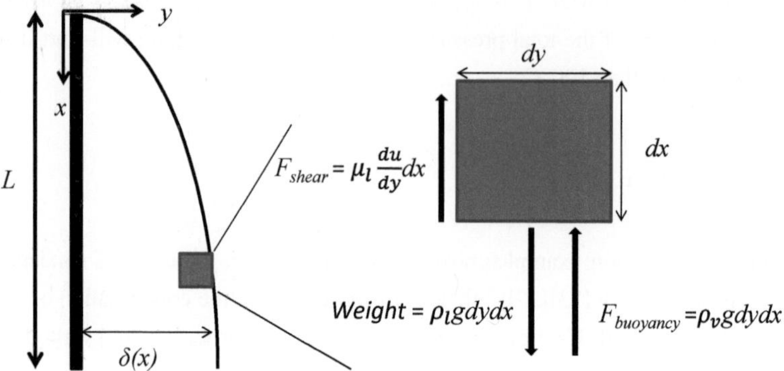

Figure 2-5. Liquid film of condensed vapour with force balance on fluid element, for a flat plate with unit width.

By integrating this expression over the film thickness and applying an energy balance on the differential element, the local Nusselt number (the Nusselt number at a certain location x) is obtained:

$$\mathrm{Nu}_x = \frac{h_l x}{\lambda_l} = \left(\frac{\rho_l (\rho_l - \rho_v) g \tilde{L} x^3}{4 k_l \mu_l (T_{sat} - T_w)} \right)^{1/4}$$

2-44

By defining a mean heat transfer coefficient over the entire film length L a mean Nusselt number over L can be obtained:

$$\mathrm{Nu}_{L,av} = \frac{h_l L}{k_l} = 0.943 \left(\frac{\rho_l (\rho_l - \rho_v) g \tilde{L} L^3}{\lambda_l \mu_l (T_{sat} - T_w)} \right)^{1/4}$$

2-45

Modified versions of this relation exist which relax on some of the assumptions made for this analysis [23][27].

Boundary layer analysis

By using the boundary layer equations (equations 2-24, 2-25 and 2-26) and by setting $\frac{dp}{dx} = g(\rho_l - \rho_v)$ a Nusselt number relation can also be found. The analysis is quite lengthy but the result is essentially the same as in equation 2-45.

Analytical solutions for a turbulent boundary layer are hard to obtain and Nusselt numbers are therefore often determined empirically. Different relations can be found in literature. According to Isashenko [30] the local Nusselt number in turbulent boundary layers is given by:

$$\mathrm{Nu} = 0.0325\,\mathrm{Re}^{1/4}\,\mathrm{Pr}^{1/2} \quad (\mathrm{Re} \geq 400) \qquad \text{2-46}$$

where the Reynolds number is defined as $\mathrm{Re} = \dfrac{u_\infty \delta}{\nu}$.

According to Colburn [31], the average Nusselt number over a plate with length L is given by:

$$\mathrm{Nu}_{L,av} = 0.056\,\mathrm{Re}_L^{1/5}\,\mathrm{Pr}^{1/3} \quad (\mathrm{Re} \geq 2000) \qquad \text{2-47}$$

Grober et al. [32] proposed the following relation:

$$\mathrm{Nu}_{L,av} = 0.0131\,\mathrm{Re}_L^{1/3} \quad (\mathrm{Re} \geq 1400) \qquad \text{2-48}$$

2.3.4 Interfacial resistance

A fluid at its saturation conditions will have a maximum phase change rate due to resistance over the liquid-vapour interface. To get a better understanding of what this means the interface can be considered at molecular level. According to Carey [27], motion of vapour molecules plays a central role in phase change rate limitations over a liquid-vapour interface. By assuming a Maxwell velocity distribution [27] of the vapour and integrating it the vapour flux for a unit surface area can be obtained. Applying corrections for the bulk vapour motion as a result of condensation or evaporation results in the following relation for the phase change rate:

$$\dot{m}_{pch} = \sigma S_{ls}\sqrt{\frac{\overline{M}}{2\pi\overline{R}\,T_{ls}}}(\Gamma p_{sat,l} - p_v) \qquad \text{2-49}$$

where σ is the accommodation coefficient, describing the probability of a molecule actually undergoing a phase change. Here, it has been assumed that these probabilities are equal for evaporation and condensation processes, which is a common approximation. The pressure p_l^{sat} is the saturation pressure belonging to the temperature of the liquid at the liquid surface T_{ls}. Furthermore, because temperatures in the liquid and vapour regions are not homogenous, the temperatures are set to the temperature at the liquid vapour interface: $T_l = T_v = T_{ls}$. It is noted that this assumption is also used by FLOW Science [41]. Γ is a factor correcting for the bulk gas motion, $\Gamma = f\left(\dfrac{w_v}{\sqrt{\dfrac{2\bar{R}T_v}{M}}}\right) = f(a)$. To apply this equation to the experiments it is necessary to determine the value of a. In chapter 3 maximum phase change rates will be determined from experimental data. Results are given in appendix III. Here it can be seen that for experiment ap160 the maximum phase change rate is 4.4E-4 kg/s. Using $\dot{m}_{pch} = \rho_v S_{ls} w_v$ and setting $\rho_v = 3.2$ kg m^{-3} (at $T_{v,av} = 168$ K and $p = 160$ kPa) and $S_{ls} = \pi R^2 = 0.066$ m^2 the vapour velocity (vapour flux) is obtained, $w_v = 2.1$E-3 m/s. The value of a is then obtained by inserting $T_{v,av} = 168$ K (appendix III) and $\bar{M} = 0.028$ kg/mol. This yields $a = 6.6$E-6. Figure 2-6 shows Γ as a function of a. As can be seen, for small a $\Gamma = 1$.

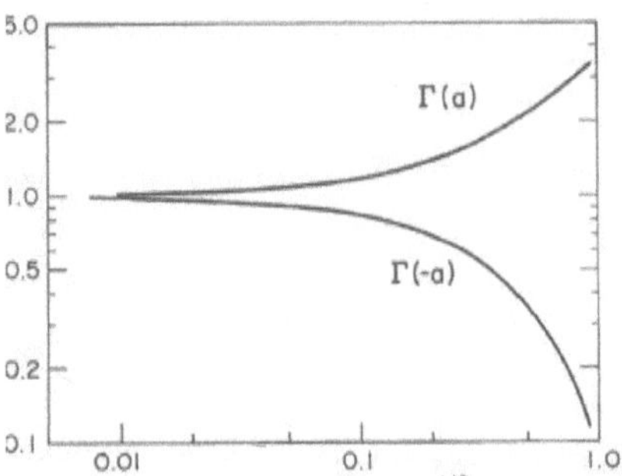

Figure 2-6. The functions $\Gamma(a)$ and $\Gamma(-a)$ [27].

2.4   Sloshing

2.4.1   Mechanics of sloshing

The response of liquids enclosed in a container depends on frequency and amplitude of the excitation and geometry of the container. Here only lateral excitation will be considered, because this is where most of this research was focussed on. Only circular cylindrical tanks are considered because this tank geometry is very common in launch vehicle applications. Also the experiments executed at ZARM were done in a circular cylindrical dewar.

Liquid oscillations in a container can be described by solving the Laplace equations of free surface in a container [5] [6]. By assuming

- The tank is inelastic
- The liquid is incompressible
- Small displacements and velocities
- Irrotational flow ($\nabla \times \mathbf{u} = 0$)
- Inviscid liquid
- Surface tension is negligible, because gravitational forces are dominant (Bo >> 1)

the Laplace equation (in cylindrical coordinates) results:

$$\frac{\partial^2 \phi}{\partial r^2} + \frac{1}{r}\frac{\partial \phi}{\partial r} + \frac{1}{r^2}\frac{\partial^2 \phi}{\partial \theta^2} + \frac{\partial^2 \phi}{\partial z^2} = 0 \qquad\qquad 2\text{-}50$$

With the following boundary conditions for a circular cylindrical tank the Laplace equation can be solved:

- The velocity normal to the tank wall is zero
- The velocity normal to the tank bottom is zero
- At the free surface the boundary condition is of the form (for derivation see [5]):

$$\frac{\partial^2 \phi}{\partial t^2} + g\frac{\partial \phi}{\partial z} = 0 \qquad\qquad 2\text{-}51$$

According to Abramson [5], a general form of the solution is:

$$\phi(r,\theta,z,t)= \sum_{m=1}^{\infty}\sum_{n=1}^{\infty} \Phi_{mn} e^{i\omega_{mn}t} \qquad\qquad 2\text{-}52$$

Φ_{mn} are the time independent normal modes of the velocity potential, where m is the mode and n the number of waves. The solution of Φ_{mn} was found by Bauer [14]:

$$\phi_{mn} = \Phi_{mn} e^{i\omega_{mn}t} = A_{mn} J_{mn}\left(\varepsilon_{mn}\frac{r}{R}\right)\cos(m\theta)\cosh\left(\varepsilon_{mn}\frac{z}{R}\right)e^{i\omega_{mn}t} = 0 \qquad 2\text{-}53$$

where A_{mn} is a constant depending on the initial conditions, ε_{mn} is given by the nth root of the derivative of the Bessel function (see appendix II) of the mth order ($J'_{mn}(\varepsilon_{mn})=0$). The roots are given in Table 2-2.

n	J_0'	J_1'	J_2'	J_3'
1	3.8317	1.8412	3.0542	4.2012
2	7.0156	5.3314	6.7061	8.0152
3	10.1735	8.5363	9.9695	11.3459

Table 2-2. The first three roots of the derivatives of the Bessel functions up to order three.

The eigenfrequency is given by [5]:

$$\omega_{mn}^2 = \frac{g}{R}\varepsilon_{mn}\tanh(\varepsilon_{mn}\frac{H_l}{R}) \qquad\qquad 2\text{-}54$$

It is observed that the lowest eigenfrequencies exist for low order modes. Surface displacement $\eta(r,\theta,t)$ can be calculated by integrating the derivative of the velocity potential with respect to z over time, which yields [5]:

$$\eta_{mn} = \frac{-i}{\omega_{mn}}\frac{\varepsilon_{mn}}{R}A_{mn}\sinh(\varepsilon_{mn}\frac{H_l}{R})J_m(\varepsilon_{mn}\frac{r}{R})\cos(m\theta)e^{i\omega_{mn}t} \qquad 2\text{-}55$$

The resulting surface shapes are given in Figure 2-7. For dynamical analysis of launcher stages, usually only the asymmetrical modes are of interest as these modes are the only ones which produce a resulting force on the stage. Asymmetric mode $m = 1$ with $n = 1$ is found in launcher stages when a lateral motion is introduced, for example by wind gusts. Symmetrical mode $m = 0$ with $n = 1$ is found during when axial excitation is introduced, for example during engine ignition.

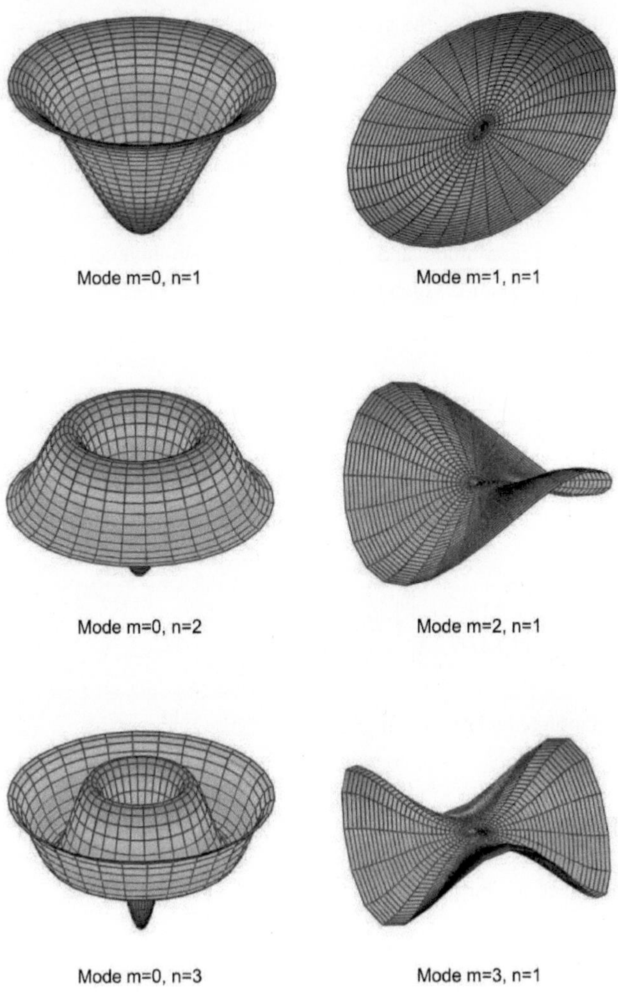

Figure 2-7. Different modes for a laterally excited liquid [28].

Damping of sloshing

Energy will be dissipated during sloshing of a liquid. The energy is dissipated at the walls and the free surface as a result of viscous boundary layers and in the liquid interior as a result of viscous stresses. In small tanks dissipation in the boundary layer dominates, in large tanks dissipation in the liquid interior. If a liquid surface is freely oscillating (no energy input), the amplitude of the slosh wave will start to decay because of the energy dissipation. This decrease is characterized by a logarithmic decrement:

$$\Delta = \ln\left(\frac{a_i}{a_{i+1}}\right) \qquad\qquad 2\text{-}56$$

where a is the wave amplitude and i is the oscillation number. A commonly found parameter is the damping ratio, which is defined as:

$$\gamma = \frac{\Delta}{2\pi} \qquad\qquad 2\text{-}57$$

For determination of the damping ratio different relations exist. For damping in a circular cylindrical container Mikishev et al. [15] derived the following relation:

$$\gamma = 0.79\sqrt{\mathrm{Re}_1}\left[1 + \frac{0.318}{\sinh\left(1.84 H_l/R\right)}\left(1 + \frac{1 - H_l/R}{\cosh(1.84R)}\right)\right] \qquad\qquad 2\text{-}58$$

for $h>2R$ the relation reduces to $\gamma = 0.79\sqrt{\mathrm{Re}_1}$

Here, Re_1 is an inverse Reynolds number defined as $\mathrm{Re}_1 = \dfrac{\nu}{\sqrt{gR^3}}$.

Stephens et al. derived the following relation [16]:

$$\gamma = 0.83\sqrt{\mathrm{Re}_1}\left[\tanh\left(1.84 H_l/R\right)\left(1 + 2\frac{1 - H_l/R}{\cosh\left(3.86\,H_l/R\right)}\right)\right] \qquad\qquad 2\text{-}59$$

when the liquid depth is large the relation reduces to $\gamma = 0.83\sqrt{\mathrm{Re}_1}$

For cylindrical containers with a spherical dome the damping increases significantly when the surface level is lower than the dome height. According to [15] the damping can be predicted by multiplying equation 2-58 with a correction factor. This correction factor can be determined from Figure 2-8.

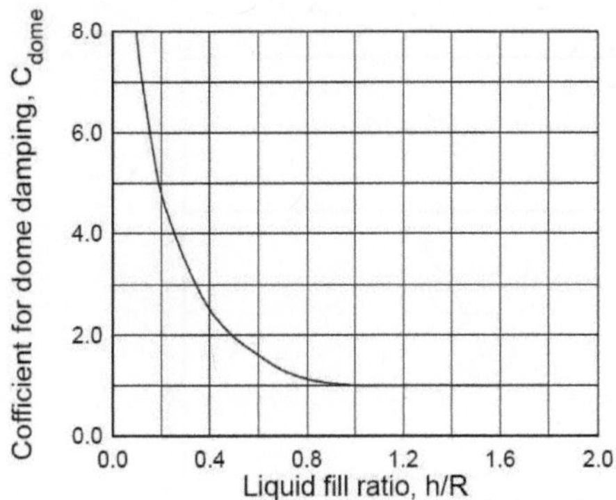

Figure 2-8. Correction factor for dome damping.

For spherical tanks the damping is given by the following relations [6]:

$$\gamma = 0.79\sqrt{Re_1}\,\frac{R}{h} \qquad\qquad \text{for } 0.1R \leq h \leq R$$

$$\gamma = 0.79\sqrt{Re_1}\,\frac{\left(1+0.46\left(2-\dfrac{h}{R}\right)\right)}{\left(1.46\left(2-\dfrac{h}{R}\right)\right)} \quad \text{for } h \geq R \qquad\qquad 2\text{-}60$$

Mechanical models of sloshing

An alternative method of determining the liquid response is the use of equivalent mechanical models [5] [6]. Mechanical models can consist of spring mass systems or pendulum systems. Here, the pendulum method will be described because it is easier to predict surface deflection using this model. The derivation of the spring mass model is however quite similar.

Liquid in a container undergoing a lateral excitation can be modelled as a mass m which remains undisturbed by a lateral motion (no motion relative to the container) and a disturbed mass modelled as a pendulum for each normal mode, see Figure 2-9. The size of the pendulum decreases rapidly for increasing mode numbers so that it is acceptable to only include the pendulum corresponding to the fundamental mode (as long as the exciting frequency is not near the eigenfrequency of a higher mode) [5]. Sloshing can be approximated by linear models when the wave amplitude does not exceed about 10% of the container diameter [6].

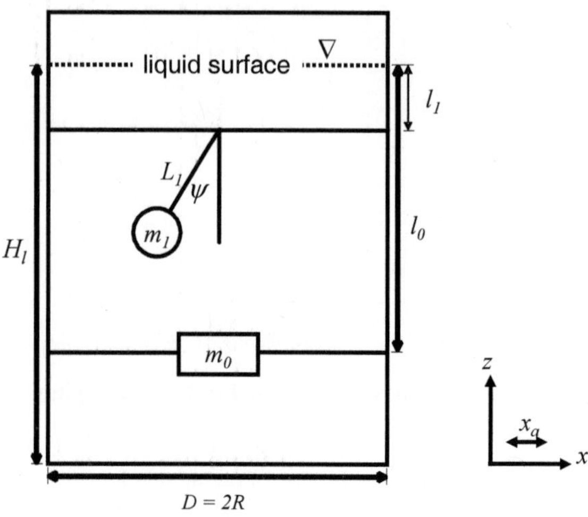

Figure 2-9: Pendulum slosh model.

The linearised equation of motion of the pendulum including damping and undergoing a harmonic lateral excitation is:

$$\ddot{\psi} + \beta\dot{\psi} + \frac{g}{L_1}\psi = -\frac{1}{L_1}x_a\omega^2\cos\omega t \qquad\qquad 2\text{-}61$$

where ψ is the deflection angle, L_1 the equivalent pendulum length and $\beta = 2\gamma\omega_{11}$ (with ω_{11} the eigenfrequency of the first mode given by equation 2-54). The equivalent pendulum length depends on the container radius and fill height, and can be determined from equations provided in [5][6]. The equations for determination of the slosh mass m_1 as well as the values for l_0 and l_1 can also be found here, and are presented in Table 2-3.

L_1	$\dfrac{D}{3.68}\coth\left(3.68\dfrac{H_1}{D}\right)$
m_1	$m_1\left(\dfrac{D}{4.4H_1}\right)\tanh\left(3.68\dfrac{H_1}{D}\right)$
m_0	$m_1 - m_1$
l_0	$\dfrac{m_1}{m_0}\left(\dfrac{H_1}{2}-\dfrac{D^2}{8H_1}\right)-(l_1+L_1)\dfrac{m_1}{m_0}$
l_1	$-\dfrac{D}{7.36}\operatorname{csch}\left(7.36\dfrac{H_1}{D}\right)$

Table 2-3. Pendulum slosh model parameters for a circular cylindrical container.

The steady state solution to equation 2-61 is:

$$\psi(t)_{ss} = \frac{-\dfrac{x_a\omega^2}{L_1}}{\sqrt{\left(\omega_{11}^2 - \omega^2\right)+\beta^2\omega^2}}\cos(\omega t).$$

$$2\text{-}62$$

$$\dot{\psi}(t)_{ss} = \frac{\dfrac{x_a}{L_1}\omega^3}{\sqrt{(\omega_{11}^2 - \omega^2)^2 + \beta^2\omega^2}}\sin(\omega t)$$

$$2\text{-}63$$

By assuming that the pendulum is always perpendicular to the liquid mass and because small surface deflections are assumed the wave amplitude can be calculated by multiplying equation 2-62 with the container radius, provided $l_1 \ll D$:

$$a(t)_{ss} = \frac{-R\dfrac{x_a}{L_1}\omega^2}{\sqrt{(\omega_{11}^2 - \omega^2)^2 + \beta^2\omega^2}}\cos(\omega t),\quad a_{\max ss} = \frac{R\dfrac{x_a}{L_1}\omega^2}{\sqrt{(\omega_{11}^2 - \omega^2)^2 + \beta^2\omega^2}}$$

$$2\text{-}64$$

Similarly, the velocity at the container wall can be calculated by multiplying equation 2-63 with the container radius:

$$u(t)_{ss} = \frac{R\frac{x_a}{L_1}\omega^3}{\sqrt{(\omega_{11}^2 - \omega^2)^2 + \beta^2\omega^2}}\sin(\omega t), \quad u_{max\,ss} = \frac{R\frac{x_a}{L_1}\omega^3}{\sqrt{(\omega_{11}^2 - \omega^2)^2 + \beta^2\omega^2}} \qquad \text{2-65}$$

2.4.2 Thermodynamics of sloshing

2.4.2.1 Thermal stratification

Thermal stratification in the liquid is a significant contributor to pressure drop effects during sloshing. It is therefore of interest to understand the physics behind the development of thermal stratification in a liquid.

Because the temperature of cryogenic propellants will usually be lower than the outside temperature of the tank containing the propellant, heat will enter the liquid. For simplicity it is assumed that heat only enters the liquid through the vertical walls. The temperature of the liquid at the tank wall will increase, giving rise to natural convective motion. The warmer liquid will be transported upwards in a natural convection boundary layer. Near the liquid surface the warm liquid will gather and a pocket of warm liquid will form. The volume of this pocket will increase over time as new warm liquid is transported upward by the boundary layer. A schematic picture is given in Figure 2-10.

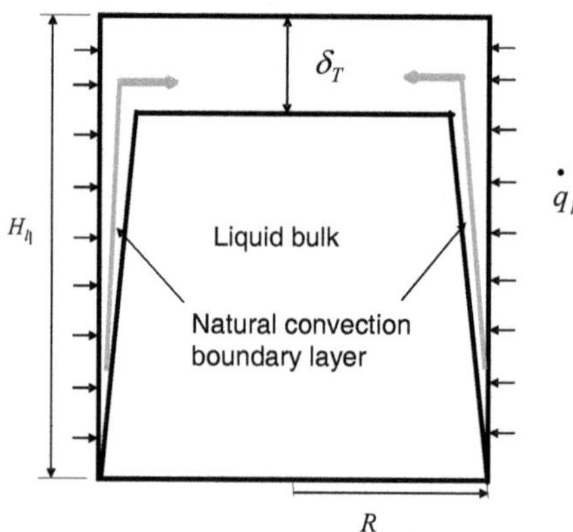

Figure 2-10. Formation of thermal stratification in a liquid due to natural convection.

The rate of growth of the stratified region can be determined by looking at the mass flow rate of liquid in the boundary layer. The mass flow rate is determined by solving the boundary layers equation according to established theory for natural convection of fluids on vertical plates. In Chin et al. [21] the following equation is analytically derived for turbulent boundary layers:

$$\delta_T(t) = H_l - H_l \left(1 + 0.0924 \frac{vt}{RH_l} \frac{1}{\mathrm{Pr}^{2/3}} \left(\frac{\mathrm{Gr}^*_{H_l}}{1 + 0.443\,\mathrm{Pr}^{2/3}} \right)^{2/7} \right)^{-7} \qquad \text{2-66}$$

Where Gr^*_{Hl} is the modified Grashof number (equation 2-30) using H_l as the characteristic length. This modified Grashof number is usually more convenient in the analysis of stratification because the heat flux is usually easier to predict/determine than wall temperature.

In Chin et al. [21] the following equation is analytically derived for laminar boundary layers:

$$\delta_T(t) = H_l - H_l \left(1 - 0.63 \frac{vt}{RH_l} \frac{\mathrm{Gr}^{*\,1/5}_{H_l}}{\mathrm{Pr}^{0.388}} \right)^5 \qquad \text{2-67}$$

Assumptions on the temperature distribution in the boundary layer together with an energy balance and assumptions on the temperature distribution in the stratified layer make it possible to determine the temperature of the stratified layer analytically. The liquid *surface* temperature is given by:

$$T_{ls} = \frac{2\left(\dfrac{H_l}{R}\right) \dfrac{vt}{H_l^2}}{I\,\mathrm{Pr}\left(\dfrac{\delta_T}{H_l}\right)} \cdot \dot{q_l} \frac{H_l}{\lambda} + T_{bulk} \qquad \text{2-68}$$

where I is a dimensionless value which describes the temperature distribution in the liquid ($I = 1$ for a homogenous temperature in the stratified layer and $I = 0.5$ for a linear temperature distribution).

According to Ring et al. [8] the growth *rate* of the stratified layer is analytically derived and yields:

$$\dot{\delta}_T = \frac{2hH_C}{R\rho c_p \xi} \qquad \text{2-69}$$

where h is the wall-liquid heat transfer coefficient (can be determined from equations in section 2.2.2), H_C is the height of the liquid at bulk temperature and ξ is a factor depending on the boundary layer profile. For turbulent boundary layers, $\xi \approx 0.25$. The thickness δ_T can be determined numerically, where H_C of course decreases over time.

The liquid surface temperature can then be determined analytically as:

$$T_{ls} = \frac{\dot{Q}_l t}{Ic_p \rho \delta_T \pi R^2} + T_{bulk} \qquad \text{2-70}$$

Thermal stratification in a liquid can also be introduced when there is energy exchange between the liquid and ullage vapour over the liquid surface. If there is heat exchange from for example hot ullage vapour to colder liquid, the liquid temperature at the surface will increase, which in turn causes a conductive heat flow into the liquid. This conductive heat flow will create a stratified layer which thickness increases over time. The thickness of the stratified layer can be determined analytically as:

$$\delta_T = C\sqrt{\alpha t} \qquad \text{2-71}$$

where α is the diffusion coefficient of the liquid and C is a constant. Das & Hopfinger propose to use $C = 3$ [20]. The thickness and temperature distribution of the stratified layer can also be determined by setting up a 1D conduction model.

Heat exchange between liquid and ullage can also take place by phase changes, where an evaporating liquid extracts thermal energy from its surroundings and condensating vapour adds thermal energy to its surroundings.

2.4.2.2   Phase change during sloshing

As mentioned in chapter 1, sloshing of cryogenic liquids can lead to thermodynamic issues (such as pressure drops) in closed containers. This aspect of sloshing has not been the subject of extensive research. Das & Hopfinger [20] are the first to publish detailed investigations on the pressure drop phenomenon. They conducted experiments using the liquids FC-72 and HFE 7000. Although not of cryogenic nature, thermodynamic environment in the system was such that pressure drop could be produced. Das & Hopfinger developed a scaling model which described pressure drop rate as a function of an effective diffusion constant. The experimentally determined dimensionless effective diffusion constant is used as a similarity parameter to scale the experimental results to full size tanks:

$$\alpha_e^* = -u_v \frac{\delta_T}{Ja} \frac{1}{\sqrt{gR^3}} \qquad\qquad 2\text{-}72$$

where u_v is the vapour flux (determined from the experimentally measured pressure drop rate), δ_T is the thickness of the thermally stratified layer in the liquid and Ja the Jacob number ($Ja = \dfrac{\rho_l c_{pl}}{\rho_v \widetilde{L}} \Delta T_l$, where $\widetilde{L}$ is the latent heat of evaporation and ΔT_l is the temperature difference over the thermally stratified layer). The effective diffusion coefficient is made dimensionless through the characteristic velocity $\sqrt{gR}$ and the characteristic length R. The experiments by Das & Hopfinger were conducted in both the stable and unstable slosh regions near the eigenfrequency of the system. Measured pressure drops showed a clear dependency on the sloshing dynamics, where the chaotic sloshing in the unstable region resulted in much higher pressure drop rates. Pressure drop rates varied from 4.2 kPa/s to 0.25 kPa/s, total pressure drop varied from 88 kPa to 18 kPa. All experiments had an initial pressure of approximately 200 kPa. Das & Hopfinger showed that in their experiments, condensation of ullage vapour during sloshing is the main mechanism for explaining the pressure drop in the system.

Pressure drop rates predicted by the model are compared to pressure drop rates according to experimental data obtained by Moran et al. and Lacapere et al., with limited success. A comparison with one of the experiments of Moran et al. resulted in a fairly good agreement. The model predicted a pressure drop rate of -160 mbar/s compared to an experimentally measured pressure drop rate of -175 mbar/s.

3   Experimental Setup and Results

This chapter will discuss the setup and the results of the experiments conducted at ZARM. These experiments are the basis for the research. In the chapters that follow, models will be developed to simulate these experiments.

3.1   Experimental setup

The experiments performed at ZARM involve a cylindrical dewar with a spherical bottom made out of borosilicate glass. The cylindrical and spherical part of the dewar have an internal radius of $R = 0.145$ m and the total height H of the dewar is 0.65 m. The space between the walls is evacuated to minimize heat transfer into the dewar by means of conduction and convection. The walls are silverized to minimize heat transfer by radiation. Thickness of the walls is 6 mm, the vacuum space between the wall has a thickness of 13 mm. A detailed schematic drawing is provided in appendix I.

A polyacetal lid was used to close the dewar, using a silicone gasket as a seal. The lid protrudes 0.015 m into the dewar, reducing the effective height to 0.635 m. A sensor boom was connected to the lid containing 15 temperature sensors and 1 pressure sensor. An additional temperature sensor was placed directly on the inside of the lid. Sensor characteristics are listed in Table 3-1.

	error	sampling rate
Pressure sensor	± 0.1 % of the measured value	100 Hz
Temperature sensor	± 22 mK (calibrated)	0.5 Hz

Table 3-1. Sensor characteristics.

The sensor boom consists of two short sensor holders containing 4 temperature sensors each. These sensors measured ullage temperatures at different heights. A longer sensor holder contained 5 temperature sensors for measurement of the liquid temperature at different heights. A picture of the sensor boom fixed to the lid is showed on the left part of Figure 3-1. The location of the temperature sensors once the lid has been placed on the dewar is provided on the right part of Figure 3-1. The liquid temperature sensors are indicated by T11, T12, T13, T14 and T15 and are located in the liquid at heights of 0.278 m, 0.253 m, 0.203 m, 0.103 m and 0.003 m respectively relative to the dewar bottom. Eight ullage temperature sensors indicated by T1 till T8 are connected to the two parallel sensor holders and are located at 0.334 m 0.384 m 0.434 m and 0.484 m with respect to the dewar bottom. One ullage temperature sensor ndicated by T9 is located at 0.55 m. At this height an aluminium baffle was installed as well. This baffle assured that when actively pressurizing the dewar the pressurization gas did not directly impinge on the liquid. Finally temperature sensor T10 is attached directly on the inner side of the lid at 0.635 m. Before putting the lid on, the dewar is filled to a height $H_l = 2R \pm 2$ mm with liquid nitrogen at ambient pressure.

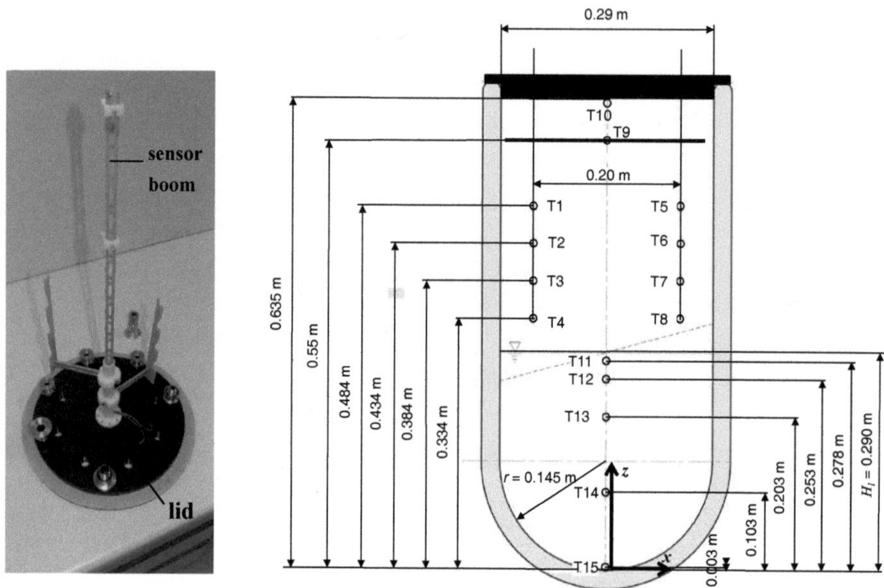

Figure 3-1. Temperature sensor boom connected to polyacetal lid (left) and schematic view of the dewar showing dimensions and sensor locations (right).

After the lid has been put on, the pressure in the system is increased to a predefined value. The dewar is pressurised by two different mechanisms. The first is by self-pressurisation and the second by active pressurisation. Self-pressurization test will be identified with "sp", followed by a number indicating the maximum pressure during pressurization. Active pressurization test will be identified with "ap", followed by a number indicating the maximum pressure during pressurization.

In case of self-pressurisation the pressure will increase due to unavoidable heat leaks from the surroundings into the closed dewar. The heat causes some of the liquid to evaporate and the pressure to rise. Heat input into the liquid can be determined using the temperature and pressure data. Total heat input during the pressurisation phase is determined to be about 7.1 W, from which 6.7 W enters the liquid and 0.4 W the vapour (see section 3.4).

Active pressurization is achieved by actively adding gaseous nitrogen. The mass flow was set to 0.2 g/s and temperature of the gaseous nitrogen was at 293 K. Active pressurization caused the pressure to increase much faster compared to the self-pressurization.

After a certain increase in pressure the dewar is exited laterally, causing the fluid to slosh. This is achieved by connecting the dewar to a crank shaft which in turn is connected to an electrical engine. A schematic view of the test setup is shown in Figure 3-2. The crank shaft has a length L_{cs} = 1.1 m. The rotational speed of the engine can be varied so that lateral sloshing at different frequencies could be obtained. The amplitude x_A of the lateral motion is given by the length of the offset of the attachment point of the crankshaft to the engine with respect to the centre of rotation of the engine. In this case this was x_A = 0.01 m. The acceleration of the dewar resulting from the lateral excitation is given by:

$$n_y = x_A\left(-\omega^2\cos(\omega t)+\frac{x_A}{L_{cs}}\omega\cos(2\omega t)\right) \approx -x_A\omega^2\cos(\omega t) \text{ because } \frac{x_A}{L_{cs}}<<1 \qquad 3\text{-}1$$

The rotational speed has been set such that f = 1.4 Hz and thus ω = 8.8 rad s^{-1}. This frequency was chosen because it is far enough from the eigenfrequency to avoid chaotic, non-linear sloshing but still introduces significant slosh wave amplitudes. The eigenfrequency of the system can be calculated with equation 2-54 which yields f_{11} = 1.8 Hz. With equation 2-64 and γ = 10^{-3} the slosh wave amplitude can be calculated yielding a = 0.03 m, which is about 10% of the dewar diameter.

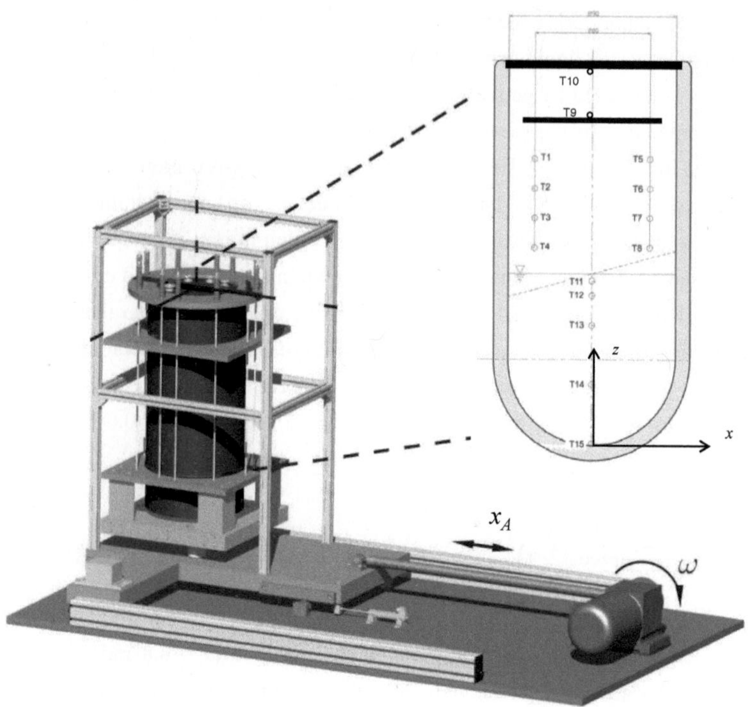

Figure 3-2. Test Setup.

3.2 Experimental Results

3.2.1 Self-pressurisation experiment results

Figure 3-3 shows the pressure measured during the self-pressurisation tests. Four different experiments were performed, where sloshing was initiated once the pressure had reached 160 kPa, 140 kPa, 130 kPa and 120 kPa. Time has been set to zero at slosh initiation. Negative times thus indicate the pressurization phase, which is longer for higher pressures. After slosh initiation, the pressure in the systems shows a short period of pressure increase. An explanation for this pressure rise might be that the cold liquid evaporates at the hotter ullage wall during the first few slosh amplitudes.

Once the ullage wall has cooled down the pressure starts to drop to a certain minimum value. After reaching this minimum value pressure slowly starts to rise again.

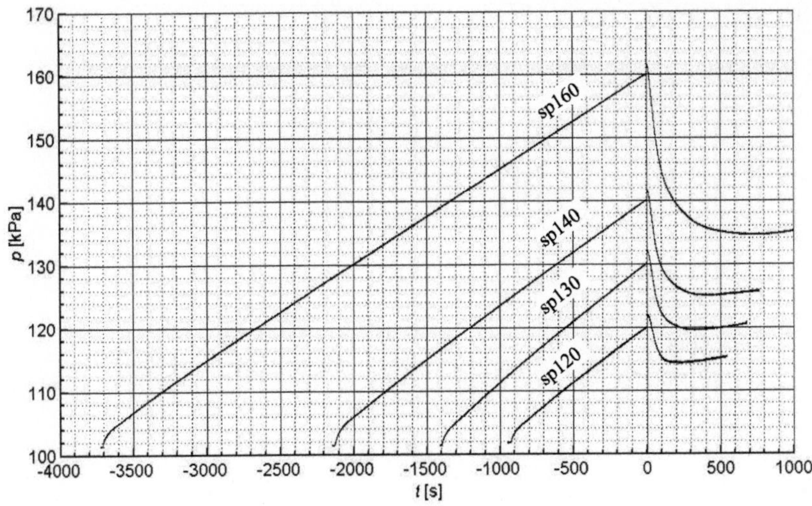

Figure 3-3. Pressure development during self-pressurisation Tests. The error corresponds to the line thickness of the graphs.

Figure 3-4 shows the pressure development after slosh initiation. In this figure, the pressure rise just after slosh initiation can be seen more clearly. The minimum pressure reached during sloshing is indicated by a symbol together with the value of the minimum pressure and the time minimum pressure is reached.

The rate of pressure change is shown in Figure 3-5. This figure has been obtained by fitting the first 100 seconds of the pressure curves in Figure 3-4 with a 5th degree polynomial. Details on the fitting function are given in appendix V. The first part of the curve where the pressure increase takes place is not taken into account in the fit as to increase the accuracy of the fit. After establishing the fit, the time derivative of the polynomial has been determined, resulting in the curves presented in Figure 3-5.

The maximum rate of pressure drop is marked with a symbol together with its value and the time it is reached. The maximum rate of pressure drop occurs within 26 to 37 seconds after slosh initiation. It is higher for experiments with higher maximum pressure and longer pressurization phases.

Characteristic data such as maximum pressure drop rate, pressurization times and minimum pressure during sloshing are given in appendix III.

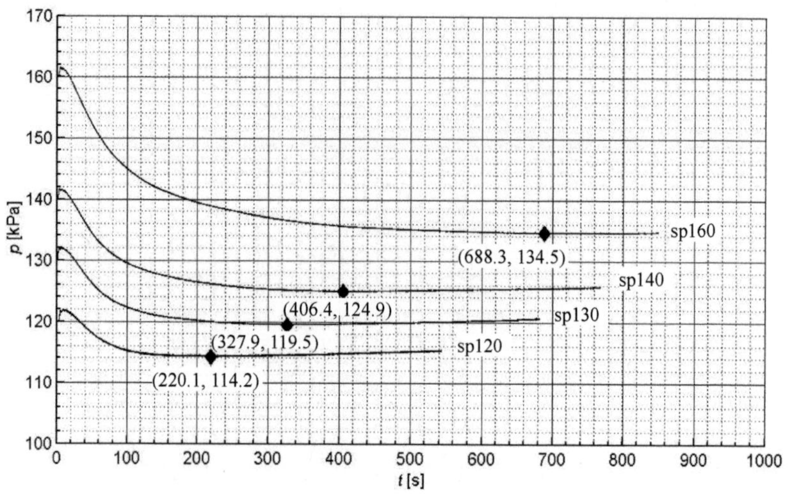

Figure 3-4. Pressure development after slosh initiation. The error corresponds to the line thickness of the graphs.

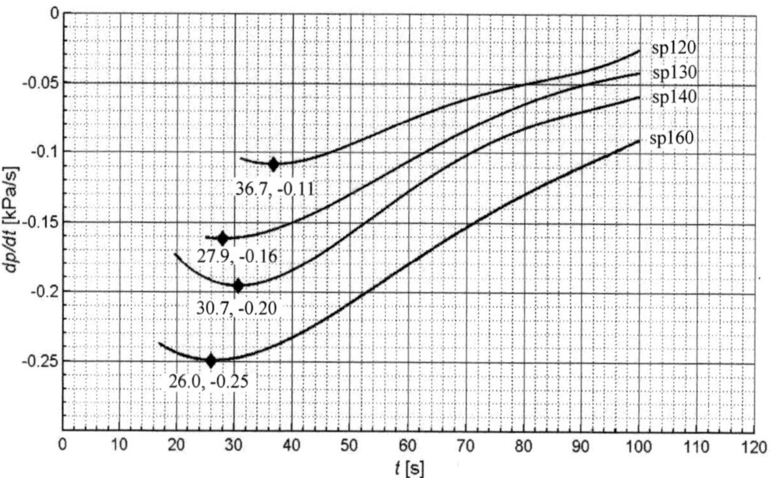

Figure 3-5. Rate of pressure change in the first 100 s after slosh initiation. The error is estimated at $\pm 5\%$.

According to the ideal gas model ($p = \rho \widetilde{R} T$) a pressure drop in a gas occurs if the gas temperature decreases, the gas density decreases or a combination of both.
The rate of pressure change is thus given by:

$$\frac{dp}{dt} = \frac{d\rho}{dt}\widetilde{R}T + \rho\widetilde{R}\frac{dT}{dt} \qquad\qquad 3\text{-}2$$

The question now arises if the temperature change, the density change or a combination of both is responsible for the pressure drop. The temperature data obtained in the experiments can be used to answer this. The temperature development in the ullage for all four self-pressurisation experiments is shown in appendix V in graphical form. Temperature data for the self-pressurized experiments are listed in tabular form in appendix III, Table III-3 to Table III-6.

By assuming a linear temperature distribution between two successive probe heights and assuming saturation temperature at $z = 0.29$ m (calculated by equation 2-42) the average temperature in the ullage, $T_{v,av}$ can be calculated (averaged over the internal energy which can be approximated by $I = mc_v T$ for gases). The specific heat c_v shows little variation with temperature and is assumed constant. The approximations can be verified using appendix VII.

The internal energy between two successive probe heights is then calculated as $I = c_v \int_{z0}^{z1} m(z)T(z)dz$). This results in the following equation for the determination of $T_{v,av}$:

$$T_{v,av} = \frac{c_v \pi R^2 \left(\int_{z=0.29}^{z=0.334m} T(z)\rho(z)dz + \int_{z=0.334}^{z=0.384m} T(z)\rho(z)dz + ... + \int_{z=0.55}^{z=0.635m} T(z)\rho(z)dz \right)}{c_v m_v} \qquad\qquad 3\text{-}3$$

where $T(z)$ is the afore mentioned linear temperature distribution between the two sensors at subsequent heights and R is the dewar radius. The density at each height can be obtained from the local temperature (according to the linear temperature distribution) and the pressure, using the ideal gas model. The gas constant $\widetilde{R}$ shows little dependency on temperature and can be assumed constant at $\widetilde{R} = 300$ J kg^{-1} K^{-1} ($\widetilde{R} = c_p\text{-}c_v$). This results in a linear density distribution between two sensor heights. The vapour mass m_v can be determined by

$$m_v = \pi R^2 \left(\int_{z=0.29}^{z=0.334m} \rho(z)dz + \int_{z=0.334}^{z=0.384m} \rho(z)dz + ... + \int_{z=0.55}^{z=0.635m} \rho(z)dz \right).$$ Vapour mass at the beginning (t_b)

and end (t_s) of the pressurization phase is listed in appendix III. The evaporated liquid mass m_{evap} is also listed in this appendix and is obtained by subtracting these vapour mass values. The average ullage temperature developments are given in Figure 3-6. During the self-pressurisation phase the average ullage temperature rises. A clear drop in the average ullage temperature is visible after slosh initiation. In Figure 3-7 the temperature development directly after slosh initiation can be seen more clearly.

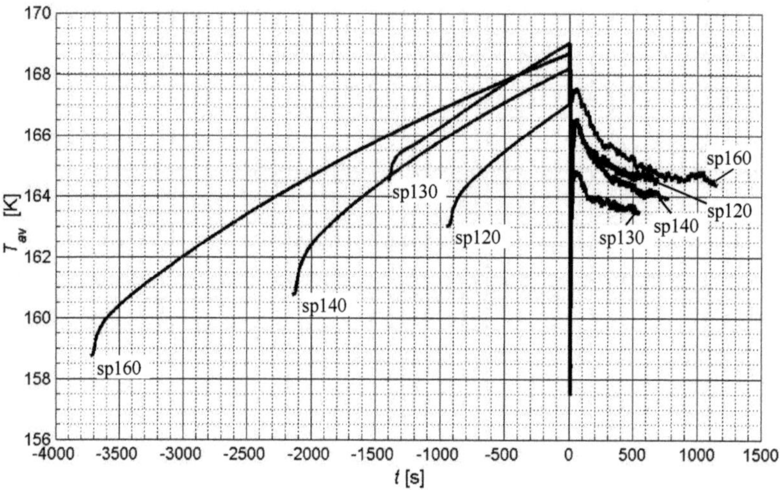

Figure 3-6. Average ullage temperature development. Error is estimated at ± 22 mK.

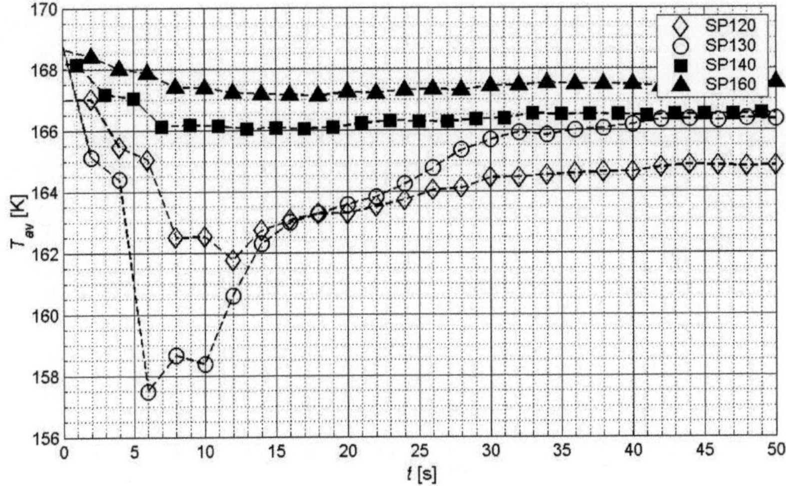

Figure 3-7. Average ullage temperature for the first 50 seconds after slosh initiation. Each datapoint is marked with a symbol. Error is estimated at ± 31 mK.

From the average ullage temperature the rate of temperature change in the ullage can be determined, and is shown in Figure 3-8 for the first 50 s after slosh initiation. In the first 10 s after slosh initiation a negative rate of temperature change occurs, indicating a temperature drop in the system. By comparing Figure 3-8 with Figure 3-4 it can be seen that this temperature drop is associated with a pressure rise. This supports the idea that during sloshing the cold liquid comes in contact with a hotter part of the ullage wall causing some of the liquid to evaporate and thus the pressure increases. As a consequence the temperature drops because cold vapour is released by the evaporation.

By comparing Figure 3-8 with Figure 3-5 again it is seen that at the point of maximum pressure drop rate, the rate of temperature change is zero. This leads to the conclusion that temperature change is not the cause for the pressure drop in the system. Rather, condensation is the main mechanism to describe pressure drop.

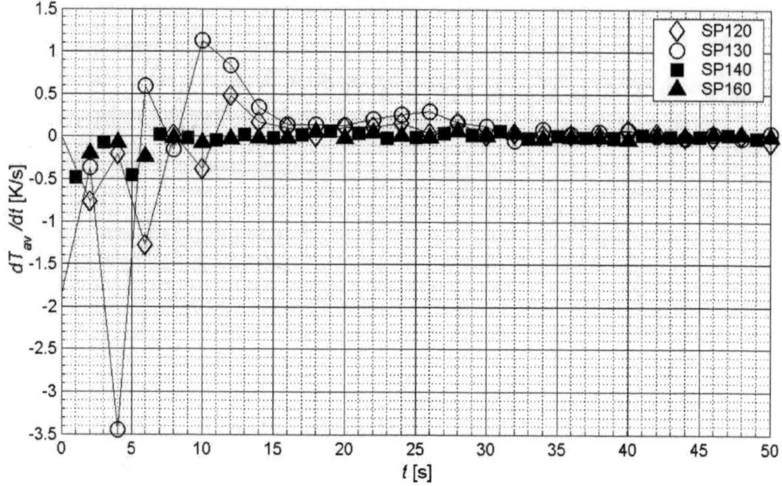

Figure 3-8. Rate of ullage temperature change for the first 50 seconds after slosh initiation. Each datapoint is marked with a symbol. Error is estimated at ± 31 mK/s.

3.2.2   Active Pressurisation Experiment Results

Figure 3-9 shows the pressure measured during the active pressurisation tests. Five different experiments were conducted, where sloshing was initiated once the pressure had reached 160 kPa, 150 kPa, 140 kPa, 130 kPa and 120 kPa. Again, time has been set to zero at slosh initiation. Negative times thus indicate the pressurization phase, which is longer for higher pressures. After slosh initiation, the pressure in the systems drops to a certain minimum value which is indicated in the figure by a symbol together with the value of the minimum pressure and the time it occurs. After the minimum pressure has been reached it slowly starts to rise again.

The rate of pressure change is shown in Figure 3-10. This figure has been obtained by polynomial fitting of the 4th degree of the pressure curves in Figure 3-9 for the first 50 seconds after slosh initiation. Details on the fitting function are given in appendix V. The first few seconds of the pressure development are not taken into account in the fit because pressure development is rather chaotic in this region. After establishing the fit, the time derivatives of the polynomials have been determined, resulting in the curves presented in Figure 3-10. The maximum rates of pressure drop are marked with a symbol together with its value and the time it is reached.

The maximum rate of pressure drop is higher for experiments with higher maximum pressure and longer pressurization phases. Also maximum rates of pressure drop are higher compared to the self-pressurization experiments.

Characteristic data such as maximum pressure drop rate, pressurization times and minimum pressure during sloshing are listed in appendix III.

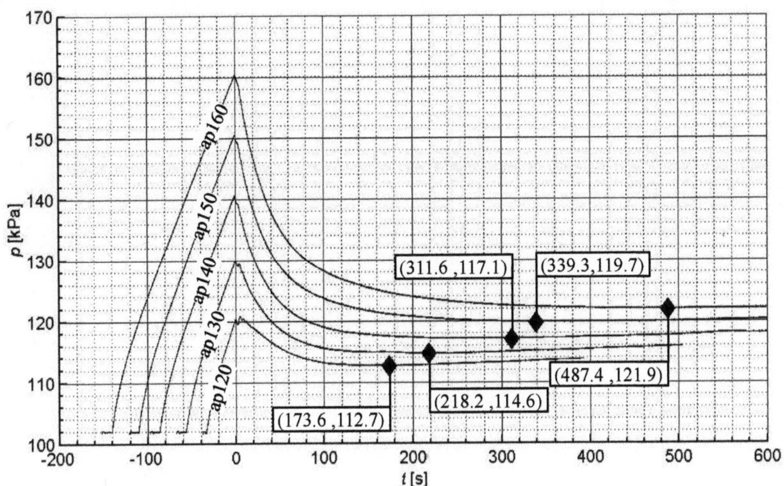

Figure 3-9. Pressure development during active pressurization tests. The error corresponds to the line thickness of the graphs.

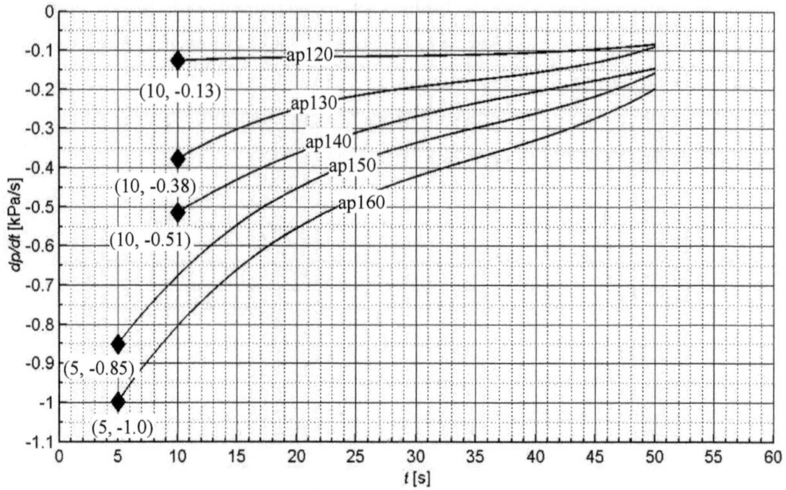

Figure 3-10. Rates of pressure change in the first 50 s after slosh initiation. The error is estimated at ± 5%.

Temperature development in the ullage for all five active pressurization experiments is depicted in graphical form in appendix V. Temperature data for the actively pressurized experiments are listed in tabular form in appendix III, Table III-7 to Table III-11.

By assuming a linear temperature distribution between two successive probe heights and assuming saturation temperature at $z = 0.29$ m (calculated with equation 2-42) the average temperature in the ullage can be calculated using the same procedure as has been done for the self-pressurized experiments. The results are provided in Figure 3-11. During the pressurisation phase the average ullage temperature rises. A clear drop in the average ullage temperature is visible after slosh initiation. This temperature drop can be seen more clearly in Figure 3-12 , where the average ullage temperature within the first 50 seconds after slosh initiation is depicted.

From the average ullage temperature the rate of temperature change in the ullage can be determined, which is shown in Figure 3-13 for the first 50 s after slosh initiation. In the first 10 s after slosh initiation a strong negative rate of temperature change occurs, indicating a temperature drop in the system.

By comparing Figure 3-13 with Figure 3-10 it is seen that the strong rate of temperature change occurs at the same time as the maximum rate of pressure drop. So in contrary to the self-pressurized experiments, temperature change has an influence on the pressure drop.

A quantitative analysis has to be done to see exactly how big the impact of the temperature change on the pressure drop rate is. Looking at ap160, it is seen that at the point of maximum rate of pressure drop, the rate of temperature drop is about -0.11 K/s. The pressure drop associated with this rate of temperature change can be calculated using equation 3-2 (where the density ρ is calculated from the maximum pressure in the system and the average ullage temperature at maximum pressure, and $\tilde{R}$ is about 300 J kg^{-1} K^{-1} at the average ullage temperature, yielding ρ = 3.2 kg/m^3). This results in dp/dt = -0.1 kPa/s, which is 10% of the total measured pressure drop rate of -1.0 kPa/s. The contributions of the rate of temperature change on the rate of pressure change for each experiment are listed in Table 3-2.

experiment	dT/dt [K/s] at max dp/dt	$T_{v,av}$ at slosh initiation [K]	ρ_{av} [kg/m^3]	dp/dt [kPa/s] due to dT/dt	% of measured dp/dt
ap120	-0.14	171	2.3	-0.10	76**
ap130	-0.1	169	2.6	-0.08	20
ap140	-0.1	168	2.8	-0.08	16
ap150	-0.11	168	3.0	-0.10	12
ap160	-0.11	168	3.2	-0.10	10

**results not reliable

Table 3-2. The contribution of the rate of temperature change on the rate of pressure change.

From this table it is clear that also in case of the active pressurization experiments condensation has to be the main mechanism describing the pressure drop.

The results for experiment ap120 form an exception in the list and do not support this conclusion. In this case the temperature change seems to be the main driver for the pressure drop. However, the results for this experiment must be treated with caution. As can be seen in Figure 3-13 the rate of temperature change fluctuates strongly for this experiment. After 12 seconds it even reaches significant positive values, although the rate of pressure change has hardly changed and is still strongly negative (compare with Figure 3-10).

Table 3-2 also shows that for experiments involving lower pressure, the relative contribution of the temperate changes are larger.

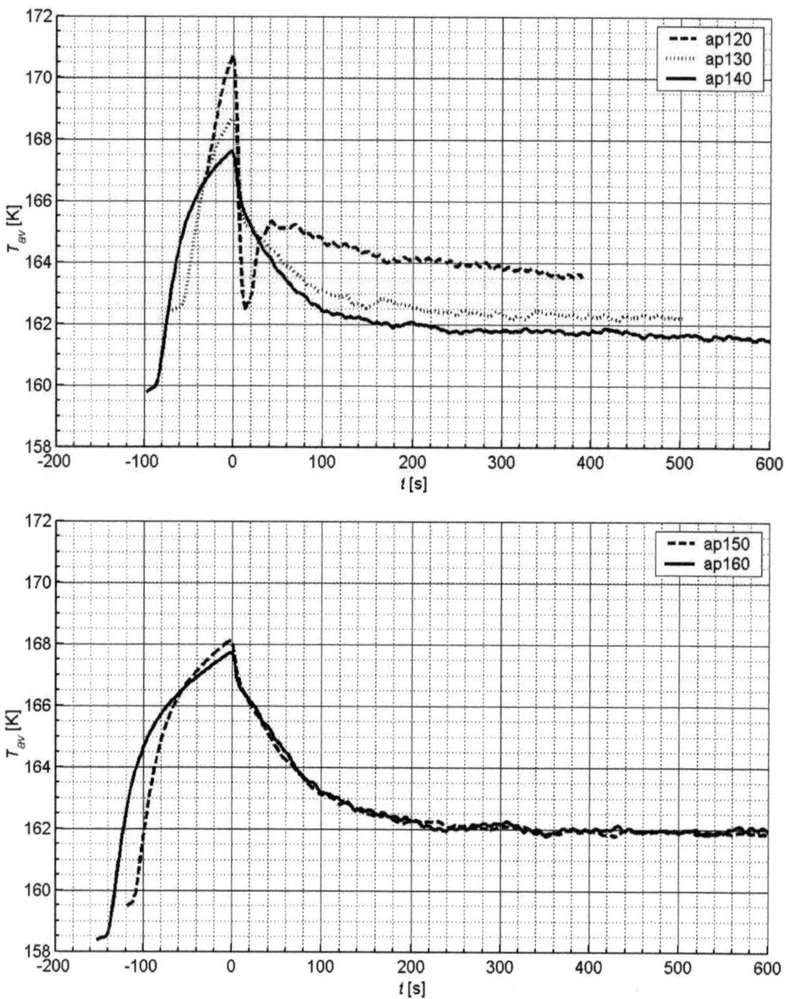

Figure 3-11. Average ullage temperature for the active pressurization experiments. Error is estimated at ± 22 mK.

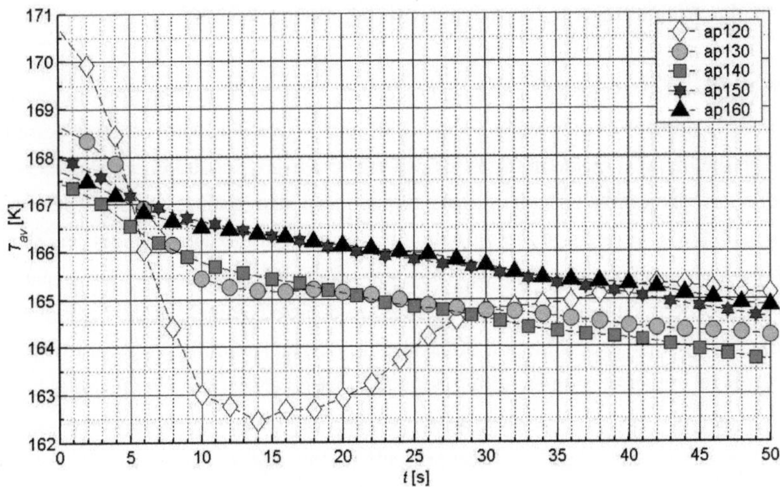

Figure 3-12. Average ullage within the first 50 seconds after slosh initiation.. Each datapoint is marked with a symbol. Error is estimated at ± 22 mK.

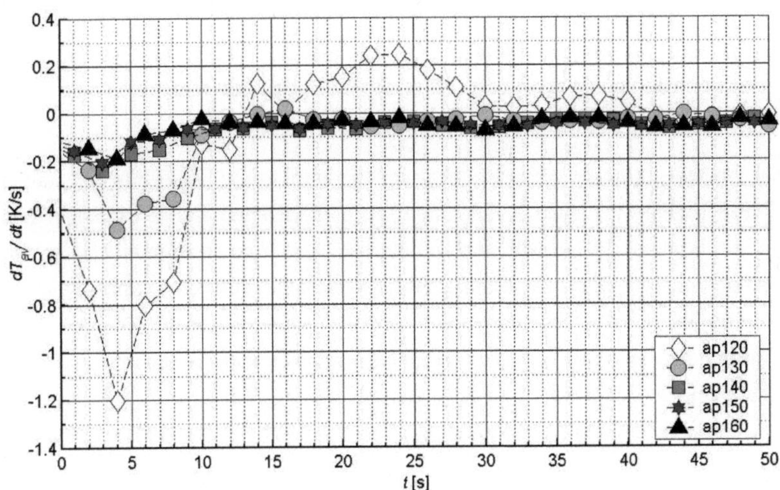

Figure 3-13. Rate of ullage temperature change withing the first 50 seconds after slosh initiation. Each datapoint is marked with a symbol. Error is estimated at ± 31 mK/s.

3.3   An explanation of the condensation occurring during sloshing

An explanation of the condensation during sloshing can be found in looking at the temperature development in the liquid. Figure 3-14 and Figure 3-15 show the temperature distribution in the liquid at different times for two of the self-pressurisation experiments and two of the active pressurisation experiments respectively. The temperatures at the liquid surface are assumed to be equal to the saturation temperature at the corresponding pressure. This saturation temperature is calculated using the Clausius-Clapyron equation (equation 2-42). For each experiment three instances in time are depicted, namely at the beginning of the pressurization phase (t_b), at initiation of the sloshing (t_s) and at minimum pressure during sloshing (t_{pmin}). At initiation of the sloshing large thermal gradients exist in the upper liquid layer. The sloshing results in a destruction of the stratification in this region and consequently a temperature drop in the upper part of the liquid. The temperature at the liquid surface is reduced, and thermodynamic equilibrium is broken. The ullage pressure is higher than the saturation pressure belonging to the temperature at the liquid surface and this leads to condensation of ullage vapor resulting in a pressure drop.

This condensation procedure is reflected by equation 2-49. Because $p_{sat,l}$ is a function of temperature it will drop as temperature decreases. Once the thermal equilibrium is broken the vapour pressure p_v will be higher than $p_{sat,l}$. This will result in a negative $\dot{m}_{pch}$ which means condensation will occur.

The thermal gradient in the liquid causes heat conduction into the liquid. For the actively pressurized experiments this means that after the pressurization has been stopped the temperature at the liquid surface will start to decrease even if no slosh motion is present. The decreasing liquid temperature will initiate the condensation procedure explained above. So even if the fluid remains at rest a pressure drop will occur as soon as the pressurization is stopped. Unfortunately this was not investigated experimentally and as such the magnitude of this pressure drop is not known.

Of course, in case of the self-pressurization experiments the pressurization cannot be stopped easily as it is caused by unavoidable heat leaks entering from the surroundings into the dewar.

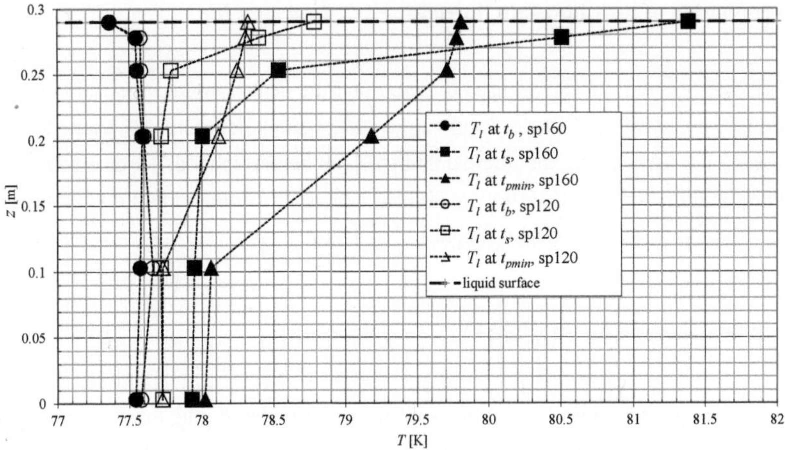

Figure 3-14. Temperature in the liquid for the Self-Pressurization tests.

The temperature values are interpolated by a linear function represented by a dotted line to ease interpretation of the figure. However it does not necessarily mean that temperature distribution between the sensor values is linear.

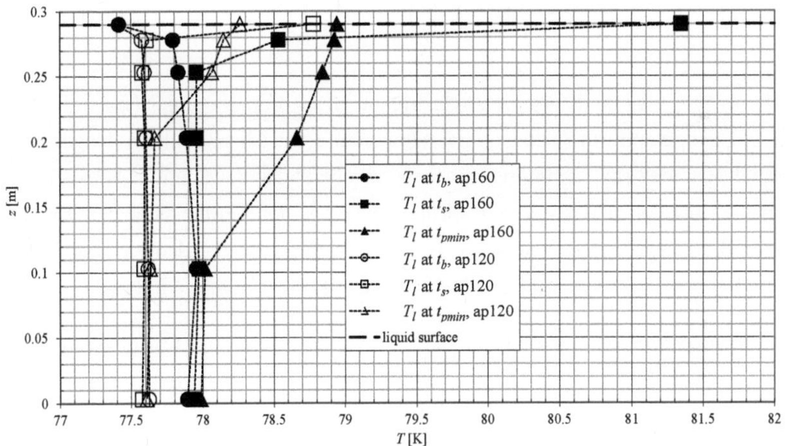

Figure 3-15. Temperature in the liquid for the Active Pressurization tests.

The temperature values are interpolated by a linear function represented by a dotted line to ease interpretation of the figure. However it does not necessarily mean that temperature distribution between the sensor values is linear.

3.4 Determination of heat flow into the fluid

Pressure and temperature data can be used to determine the heat flow into the fluid during pressurization. By assuming a linear temperature distribution between the temperatures measured during the experiment, the internal energy I can be determined. To determine the heat flow into the fluid, the heat flow will be divided in heat flow into the liquid volume and heat flow entering the ullage (vapour) volume.

3.4.1 Heat flow into the liquid volume

Because liquid evaporates during the pressurization phase the liquid volume must be modelled as an open system. The heat entering the liquid can be determined using equation 2-14. The work done on the control volume is zero during pressurization. In differential form this can be written as $\Delta I_{cv} = \Delta Q_{cv} - \Delta m_e h_e$. Over the complete pressurization phase this yields for the heat flow into the liquid: $\Delta Q_{cv} = \left(m_l - m_{evap}\right)i_{cv}\big|_{ts} - m_l i_{cv}\big|_{tb} + m_{evap}h_e$. Rearranging terms yields:

$$\Delta Q_{cv} = \Delta Q_l = m_l\left(i_{cv}\big|_{ts} - i_{cv}\big|_{tb}\right) + m_{evap}\left(h_e - i_{cv}\big|_{ts}\right)^1 \qquad\qquad 3\text{-}4$$

The internal energy i_{cv} can be determined by using the NIST property tables (appendix VII) at pressure and temperature at t_b and t_s. The temperature used here is the average liquid temperature, determined by assuming a linear temperature distribution between the temperatures measured at different sensor locations. This is done in the same sense as has been done for the ullage vapour (equation 3-3), but in this case the density shows little variation with temperature and can be assumed constant. Furthermore the internal energy in a liquid must be calculated as $I = c_p \int_{z0}^{z1} m(z)T(z)dz + I_0$, where care must be taken to use c_p instead of c_v (see discussion in appendix IV) and I_0 is the internal energy at $T(z{=}0)$. This results in the following equation allowing determination of the average liquid temperature:

---

[1] It is noted that this last term represents the latent heat

$$T_{l,av} = \frac{c_p \pi \rho_l \left(\int_{z=0}^{z=0.003\,m} T(z)r(z)^2\,dz + \int_{z=0.003}^{z=0.103\,m} T(z)r(z)^2\,dz + ... + \int_{z=0.278}^{z=0.29\,m} T(z)r(z)^2\,dz \right)}{c_p m_l} \qquad 3\text{-}5$$

where $T(z)$ is the afore mentioned linear temperature distribution between the two subsequent sensor heights and $r(z)$ is the radius of the dewar at height z (measured from the symmetry axis of the dewar to the dewar wall). This is not constant because of the dome section at the dewar bottom.

Using $\rho_l = 805$ kg/m^3, the average liquid temperatures are obtained. Based on the geometry provided in section 3.1 and taking into account the uncertainty in the fill level the liquid mass is determined to be $m_l = 12.85 \pm 0.1$ kg.

The procedure described above will be demonstrated using sp160 as an example. In section 3.2.1 it has been shown how m_{evap} can be determined. For sp160 $m_{evap} = 0.0228$ kg. Equation 3-5 is solved numerically and results in $T_{l,av}\big|_{ts} = 78.38$ K and $T_{l,av}\big|_{tb} = 77.6$ K. By looking at the NIST property tables in appendix VII it is seen that $i_{cv}\big|_{ts} = -120082$ J kg^{-1} and $i_{cv}\big|_{tb} = -121676$ J kg^{-1}. Internal energy and enthalpy are mainly a function of temperature and show minor dependency on pressure. For h_e an average value is therefore used, although pressure varies over the pressurization phase. Here $h_e = 78600$ J kg^{-1}.

Taking into account uncertainties in m_l and m_{evap} (see Table III-1) this yields $\Delta Q_l = 25013 \pm 199$ W . By dividing this through the pressurization time this results in a heat flow $\dot{Q}_l = 6.7 \pm 0.05$ W.

3.4.2 Heat flow into the ullage volume

Because evaporated liquid enters the ullage it must be modelled as an open system. Using the same procedure used to derive equation 3-4 , equation 2-14 can be written as:

$$\Delta Q_{cv} = \Delta Q_v = m_v \left(i_{cv}\big|_{ts} - i_{cv}\big|_{tb} \right) + m_{evap} \left(i_{cv}\big|_{ts} - h_i \right) \qquad 3\text{-}6$$

The vapour mass m_v at t_b has been determined in section 3.2.1 and is listed in Table III-1. Here $h_i = 78600$ J kg^{-1}. Evaluating the internal energy at t_s and t_b (using appendix VII and the average ullage temperature at t_s and t_b), $\Delta Q_{cv} = 1413$ J is obtained. This results in a heat flow into the vapour of $\dot{Q}_v = 0.73 \pm 0.01$ W. The total heat flow into the dewar for sp160 is thus 7.4 ± 0.06 W.

3.4.3   Heat flow during other self-pressurization experiments

It is expected that the heat flow is the same for the other self-pressurization experiments because the experimental conditions remained unchanged. Nevertheless, the heat flow is determined for each self-pressurization experiment to prove this assumption. The results are given in Table 3-3. The total heat flow into the dewar lies around 7 W for all experiments. Only experiment sp130 has a somewhat higher heat flow of 7.8 W.

experiment	t_p [s]	$T_l^{av}\big|_{tb}$ [K]	$T_l^{av}\big|_{ts}$ [K]	$T_v^{av}\big|_{tb}$ [K]	$T_v^{av}\big|_{ts}$ [K]	$\dot{Q}_l$ [W]	$\dot{Q}_v$ [W]	$\dot{Q}_{tot}$ [W]
SP160	3720	77.60	78.38	159	169	6.7	0.7	7.4
SP140	2143	77.65	78.05	165	169	6.5	0.4	6.9
SP130	1408	77.58	77.88	161	168	7.2	0.6	7.8
SP120	946	77.59	77.77	163	167	6.6	0.5	7.1

Table 3-3. Heat flow into the dewar during self pressurization experiments.

4   Numerical modeling of thermal stratification

As discussed in chapter 3, thermal stratification develops in the fluid during the pressurization phase. It is argued that thermal stratification in the liquid is the main cause for the pressure drop observed during sloshing.

Therefore, any attempt at simulating the observed pressure effects must include proper representation of the thermal stratification process. This chapter focuses on the numerical simulation of thermal stratification using both a full 3D commercial model, CFD code FLOW 3D [40], and a simpler 1D model which is derived partially from the full 3D results. This chapter will mainly discuss stratification during the self-pressurization experiments. Modeling this form of pressurization is the most complex, as it involves different heat flow mechanisms. The model can then easily be adapted for active pressurization.

Once the thermal stratification has been satisfactorily modeled it will be possible to model the effects of sloshing. This will be discussed in chapter 5.

4.1   Analysis of heat flow into the liquid

To be able to model the development of thermal stratification in the self-pressurisation experiments it is necessary to understand in which way heat enters the liquid. The experimental measurements of experiment sp160 (see chapter 3) are used to analyze this. Figure 4-1 shows the temperature distribution in the dewar measured at the beginning and end of the pressurisation phase. The ullage temperature shows a roughly linear distribution, where the temperature at the lid is constant at about 284 K temperature and the temperature at the liquid surface will be equal to the saturation temperature. It is therefore expected that convective flows in the vapour are limited (in case of strong convective currents a more homogeneous temperature distribution could be expected) and that heat transport through the vapour into the liquid is mainly caused by conduction.

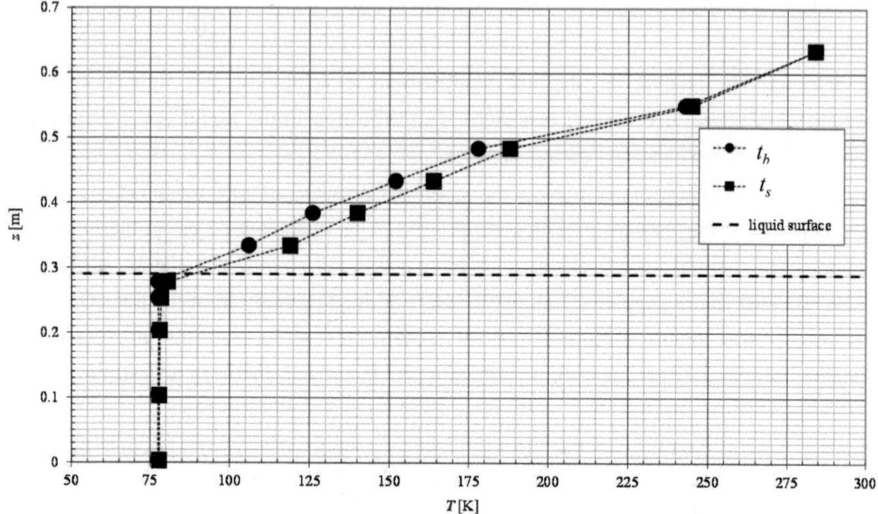

Figure 4-1. Temperature distribution in the dewar at the beginning (t_b) and end (t_s) of the pressurisation phase for sp160.

The thermal conduction coefficient of gaseous Nitrogen is very low ($\lambda_v = 0.01$ Wm^{-1}K^{-1}). Temperature gradients in the ullage near the liquid surface at t_b and t_s are estimated a 641 K/m and at 845 K/m respectively. Using the average value of this gradient (743 K/m) and equation 2-32 and multiplying with the liquid surface area the heat conducted from the vapour into the liquid is determined to be $\dot{Q}_{v-l} = 0.5$ W.

The dewar wall will have a higher temperature in the ullage region because of the high vapour temperatures. This will cause a conductive heat flow tangentially through the dewar wall from the ullage region into the liquid. Most of this heat will be released at the liquid surface, increasing the thermal gradient in the upper parts of the liquid region. A rough estimation of the heat conducted in this way can be made assuming the dewar wall has the same temperature as the nitrogen fluid. The thermal conductivity of the borosilicate glass is given in appendix VIII. As can be seen the conductivity is not constant, but varies with temperature. At the liquid surface, where temperature of the wall is assumed to be at saturation temperature, the thermal conductivity of the wall is estimated at $\lambda_w = 0.5$ Wm^{-1}K^{-1}. The surface area is given by $S = 2\pi R t$, where $t = 0.006$ m (the thickness of the wall) and R

$= 0.145$ m (the dewar radius at the liquid surface). If the temperature gradient is again assumed at 743 K/m, the conductive heat flow tangential through the wall amounts to $\dot{Q}_{w-l} = 2$ W. Combining the two factors yield a total heat flow from the ullage region into the upper part of the liquid of $\dot{Q}_{ul-l} = \dot{Q}_{w-l} + \dot{Q}_{v-l} = 2.5$ W. This is 37 % of the total heat flow into the liquid, determined at 6.7 W in section 3.4. The remaining heat will enter the liquid in normal direction through the dewar wall, as a results of unavoidable heat leaks through the insulated wall. A schematic view of the heat flow mechanisms is given in Figure 4-2, where $\dot{Q}_{l-n}$ is the remaining heat flow entering the liquid (normal to the dewar wall). During the self pressurization process, some of the liquid will evaporate as has been shown in section 3.4.1. The evaporating liquid will extract heat from the liquid at the liquid surface. This is indicated by $\dot{Q}_{pch}$.

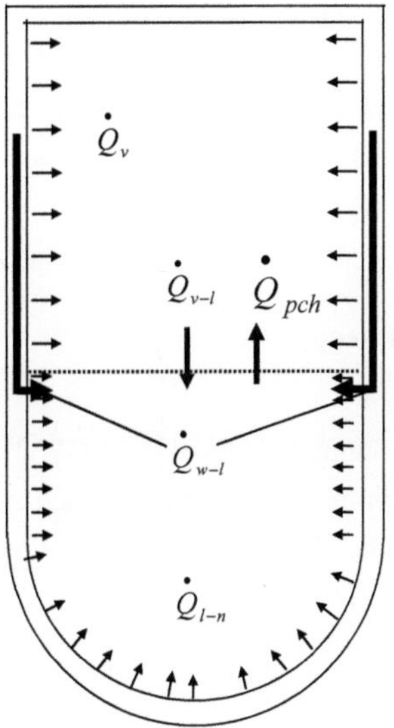

Figure 4-2. Dewar heat flow mechanisms.

As has been explained in section 2.4.2.1 the heat flux $\dot{Q}_{l-n}$ will cause the formation of a thermal boundary layer at the wall which transports warm liquid towards the surface. This boundary layer can be laminar or turbulent, where the transition from laminar to turbulent occurs at a modified Raleigh number of $Ra^*_{Lc} = 1E11$ (see section 2.2.2, $Ra^*_{Lc} = Gr^*_{Lc} Pr = Gr_{Lc} Pr\, Nu$ and the modified Grashof number Gr^*_{Lc} is defined in equation 2-30).

To determine Gr^*_{Lc}, $\lambda_l = 0.15$ W m^{-2}, $v_l = 1.98E-07$ Pa s^{-1}, $\beta = 0.0057$ K^{-1} is used which is based on appendix VII. Furthermore $g = 9.81$ m s^{-2}. The reference length L_c is the running length along the wall in contact with the liquid which is $L_c = R + 0.5\pi R$ (see geometry description in section 3.1). Finally the area specific heat flux $\dot{q}_{l-n}$ must be determined. Therefore the wall area in contact with the liquid is determined which is $S_{w-l} = 2\pi R^2 + 2\pi R^2 = 0.264$ m^2 (again see geometry description in section 3.1). The area specific heatflux is thus

$$\frac{\dot{Q}_{l-n}}{S_{w-l}} = 15.9 \text{ W/m}^2.$$

Inserting this data into equation 2-30 and multiplying with the Prandtl number (Pr = 2.2 using equation 2-38) gives $Ra^*_{Lc} = 6.4E12$. This means that laminar-turbulent transition of the boundary layer is expected. The transition length is 0.13 m (at this running length $Ra^*_{Lc} = 1E11$), but as the dewar has a spherical bottom the transition will occur sooner do to the influence of the inclined surface [23]. Therefore it is expected that the thermal boundary layer at the wall will be (almost) completely turbulent.

4.2   Mesh sensitivity analyses

Before setting up a complete numerical model in FLOW 3D, a mesh sensitivity analysis has been done to determine the minimum computational cell size required for obtaining reliable numerical results. This will be done for the two main heat flow mechanisms, namely $\dot{Q}_{w-l}$ and $\dot{Q}_{l-n}$.

In all the computations described in this section, nitrogen fluid properties are inserted in the FLOW3D model according to appendix VII for $p = 160$ kPa and $T = 77.5$ K. Initial liquid temperature distribution is always homogeneous at 77.5 K.

4.2.1 Mesh sensitivity analysis for liquid heat flow in normal direction to the dewar wall

According to theory (section 2.4.2.1) it is expected that a thermal boundary layer will form at the dewar wall which causes the heated liquid to rise along the wall. This will cause the liquid near the liquid surface to heat up. The heated volume will grow over time. A mesh sensitivity analysis is performed to determine the required cell size to capture this process. It might be necessary to resolve the thermal boundary layer at the dewar wall, which would implicate a small cell size.

As a test case a cylindrical tank is chosen with a radius $R = 0.145$ m and a height $H = R$. These dimensions correspond to liquid volume in the dewar without the spherical bottom. Disregarding the spherical bottom simplifies the analysis as thermal boundary layer formation at a vertical wall has been investigated intensively and is understood very well.

The cylinder is assumed to be heated from the sides only. $\dot{Q}_{l-n}$ has been set to 17 W, higher than actually present during the experiment. This has been done to amplify the effect, so that smaller simulated time periods yield noticeable results. This saves CPU time and makes the sensitivity study more practical. Figure 4-3 gives a schematic representation of the test case. A 2D cylindrical mesh has been used because the problem is rotationally symmetric. Five different mesh resolutions were tried: 60x60, 120x120, 240x240, 360x360 and 480x480 cells. In terms of cell size this yields 0.0024 m x 0024 m, 0.0012 m x 0.0012 m, 0.0006 m x 0.0006 m, 0.0004 m x 0.0004 m and 0.0003 m x 0.0003 m respectively.

Convergence is measured according to the growth rate of the thermally stratified layer δ_T, with respect to three points S1, S2 and S3. Point S1 is located at $x = 0.0725$ m and $z = 0.2175$ m (half way up the test section), S2 is located at $x = 0.0725$ m and $z = 0.18125$ (half way between bottom of test section and S1) and S3 is located at $x = 0.0725$ m and $z = 0.145$ (bottom of test section).

The convergence test has been done for both laminar and turbulent boundary layer settings in FLOW 3D.

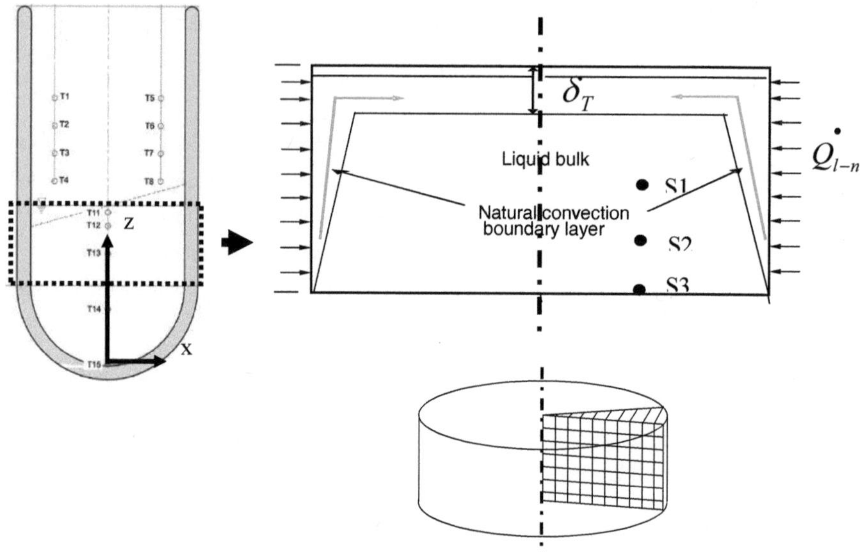

Figure 4-3. Schematic presentation of the test case used for the mesh sensitivity analysis in case of heat flow normal through the dewar wall.

Laminar boundary layer

In Figure 4-4 the temperature development over 500 s for the reference points S1, S2 and S3 is given for the different mesh resolutions and laminar boundary layer settings. A clear convergence can be seen in the figure at a resolution of 240x240 cells, because for higher resolutions the temperature development at the reference points shows equal behavior.

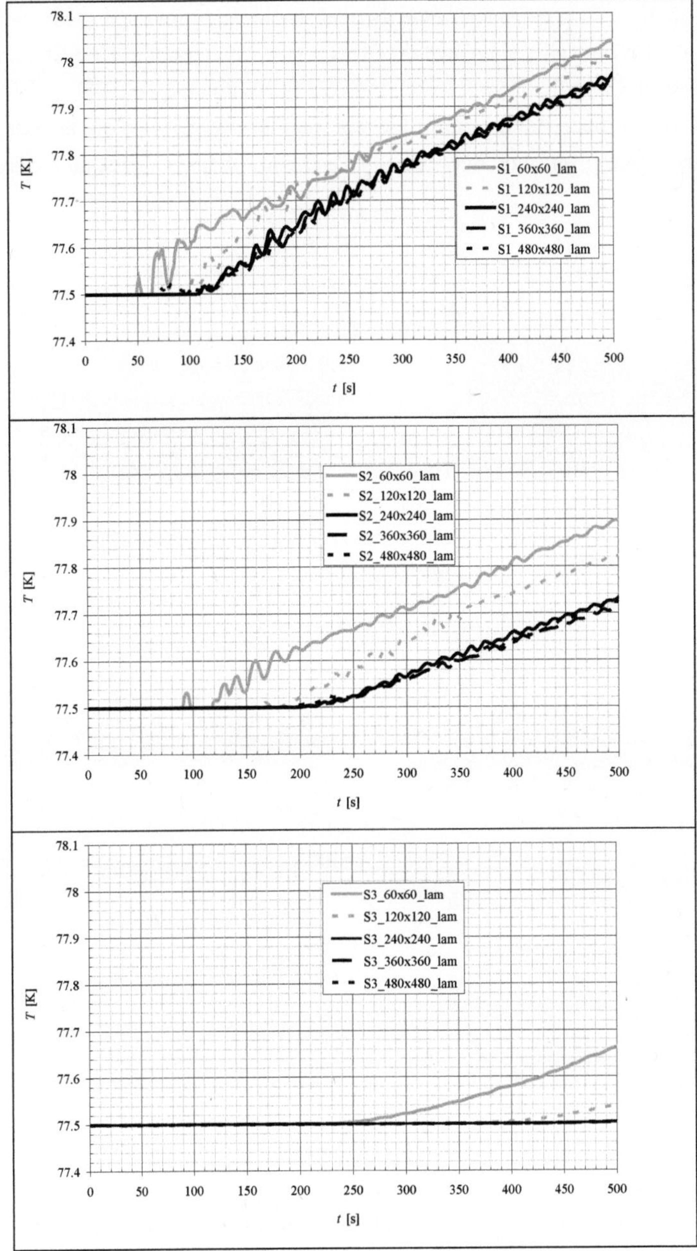

Figure 4-4. Temperature development according to FLOW 3D for reference points S1 (top), S2 (middle) and S3 (bottom).

Figure 4-5 shows the development of the thermal stratification for the test case with a mesh resolution of 240x240 cells. Simulated time was 500 s. An output for every 100 s is visible. The growth of the stratified layer can be seen clearly. Within the first 100 s, almost half of the volume is stratified. The growth of the stratified layer continues, but the rate at which this happens reduces over time. Between 400 s and 500 s there is almost no increase visible anymore.

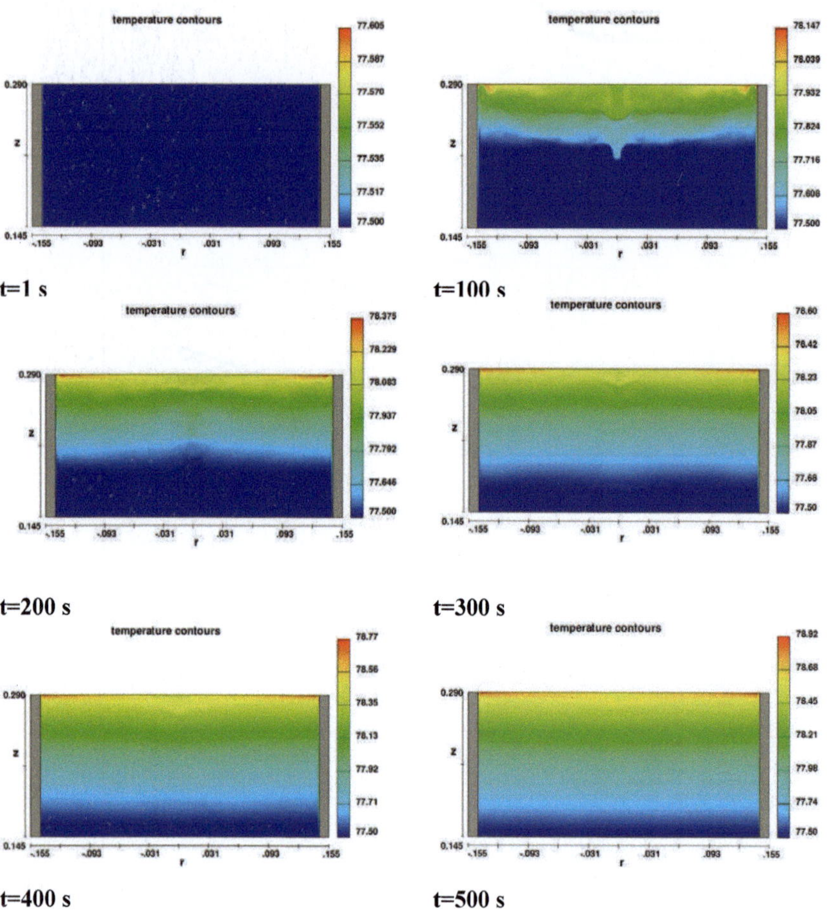

t=1 s   t=100 s

t=200 s   t=300 s

t=400 s   t=500 s

Figure 4-5. Development of stratification in the test case for a laminar boundary layer with a mesh resolution of 240x240 cells. The temperature scale is in K.

The thermal boundary layer at the wall is hardly visible in Figure 4-5 because it is very thin. The fully developed boundary layer is about 5 cells thick, which is about 0.003 m. The boundary layers for mesh resolutions of 60x60, 240x240 and 480x480 cells are compared with each other in Figure 4-6. Because the solution convergences for a mesh resolution of 240x240 cells, it can be concluded that the fully developed thermal boundary layer must be resolved by at least 5 cells.

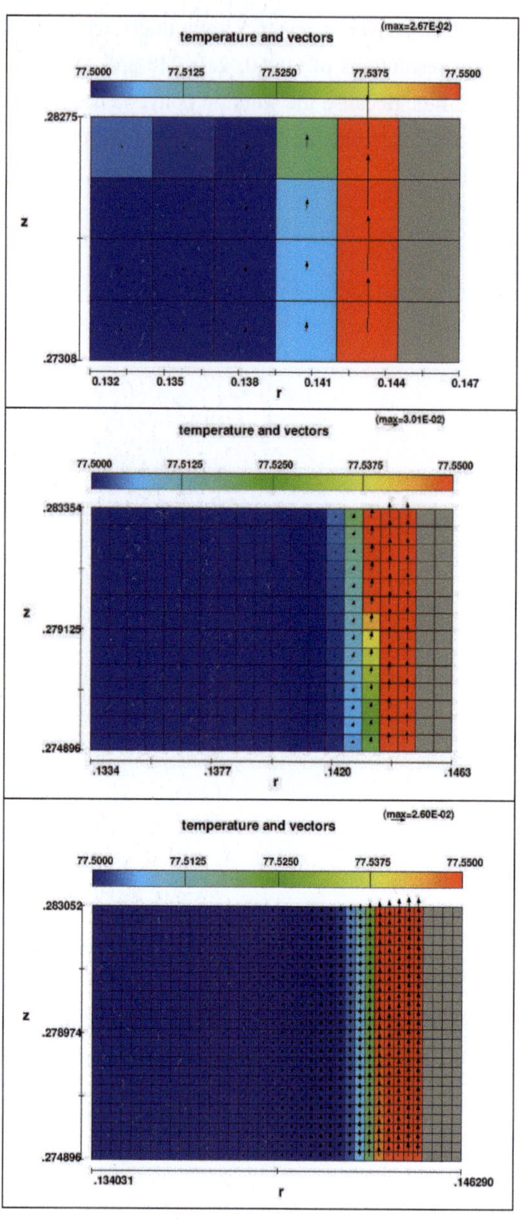

Figure 4-6. Fully developed boundary layers at t=11 s for mesh resoltions of 60x60 (top), 240x240 (middle) and 480x480 cells (bottom). Temperature scale is in K.

It is of interest to compare the development of the stratification according to FLOW 3D with the model according to Chin et al., presented in section 2.4.2.1. The results are shown in Figure 4-7. An offset between the two solutions of less than 0.02 m is present, where FLOW 3D predicts a thicker stratified layer than the model according to Chin et al. does.

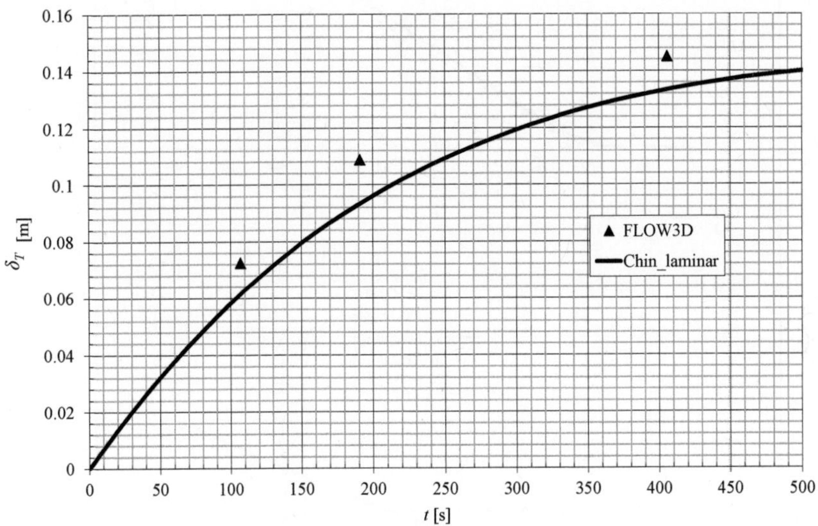

Figure 4-7. Development of stratification accoriding to FLOW3D simulations and according to Chin et al.

Turbulent boundary layer

In Figure 4-8 the temperature development over 500 s for the reference points S1, S2 and S3 is given for the different mesh resolutions and turbulent boundary layer settings. The chosen turbulence model is the k-e model. The mesh resolution required for a converged solution is the same for the turbulent case and the laminar case. A clear convergence can be seen in the figure at a resolution of 240x240 cells.

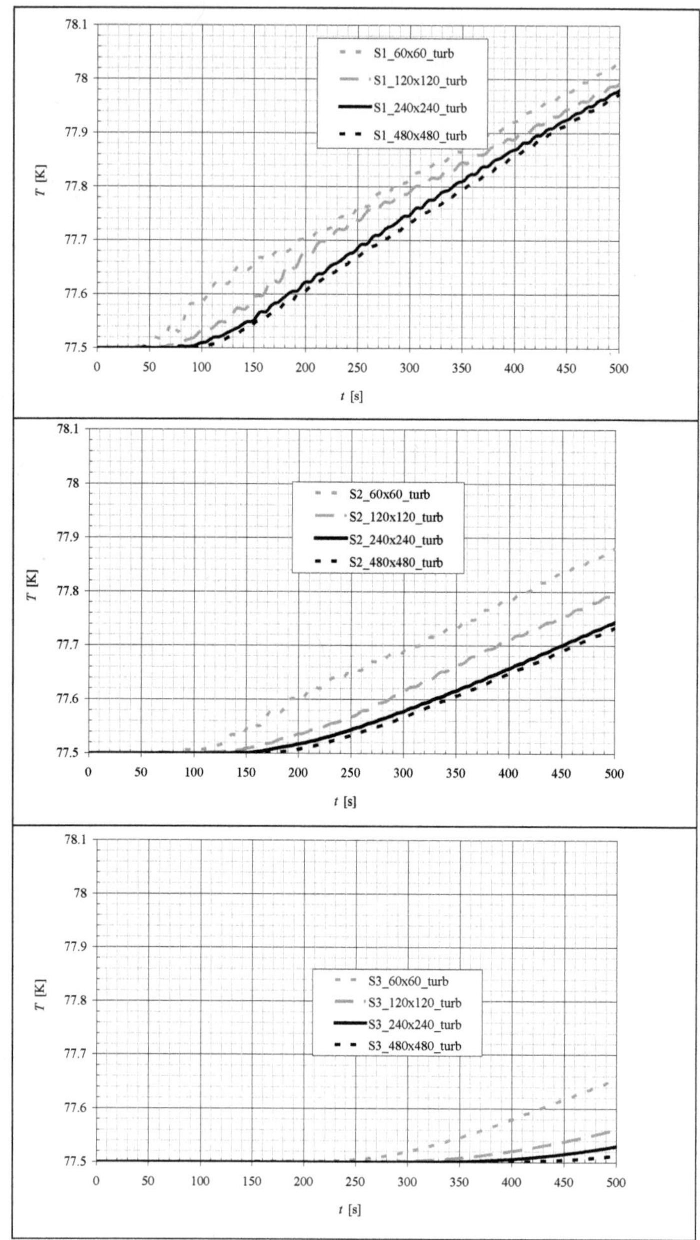

Figure 4-8. Temperature development for reference points S1 (top), S2 (middle) and S3 (bottom).

Figure 4-9 shows the development of the stratified layer for the test case with a mesh resolution of 240x240 cells. Simulated time was 500 s. An output for every 100 s is visible. The growth of the stratified layer can be seen clearly. Within the first 100 s, almost half of the volume is stratified. The growth of the stratified layer continues, but the rate at which this happens reduces over time. At 400 s the complete volume is stratified. This can be seen using the temperature scale to the right of the figure. Here it is observed that the minimum temperature in the liquid has increased from 77.5 K to 77.51 K. This is in contrast to the laminar case where the volume isn't completely stratified even after 500 s.

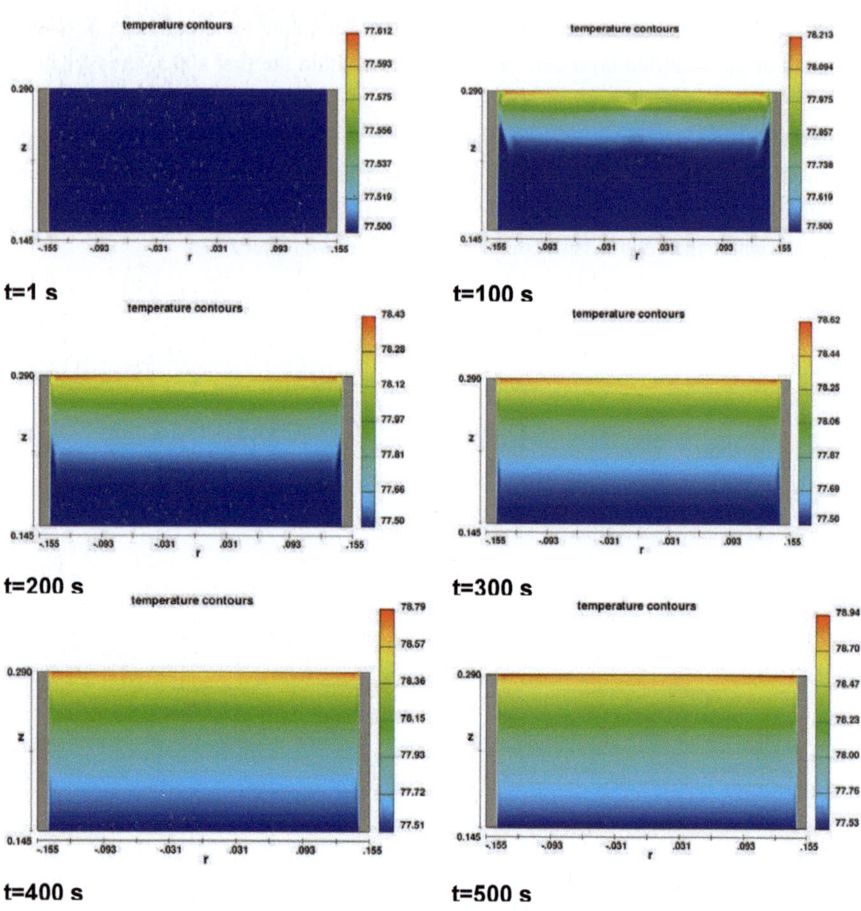

Figure 4-9. Development of stratification in the test case for a laminar boundary layer with a mesh resolution of 240x240 cells. Temperature scale is in K.

The thermal boundary layer at the wall is hardly visible in Figure 4-9 because it is very thin. The fully developed thermal boundary layer is about 5 cells thick, which is about 0.003 m. The boundary layers for mesh resolutions of 60x60, 240x240 and 480x480 cells are compared with each other in Figure 4-10. Because the solution convergences for a mesh resolution of 240x240 cells, it can be concluded that the fully developed thermal boundary layer thickness must be resolved by at least 5 cells.

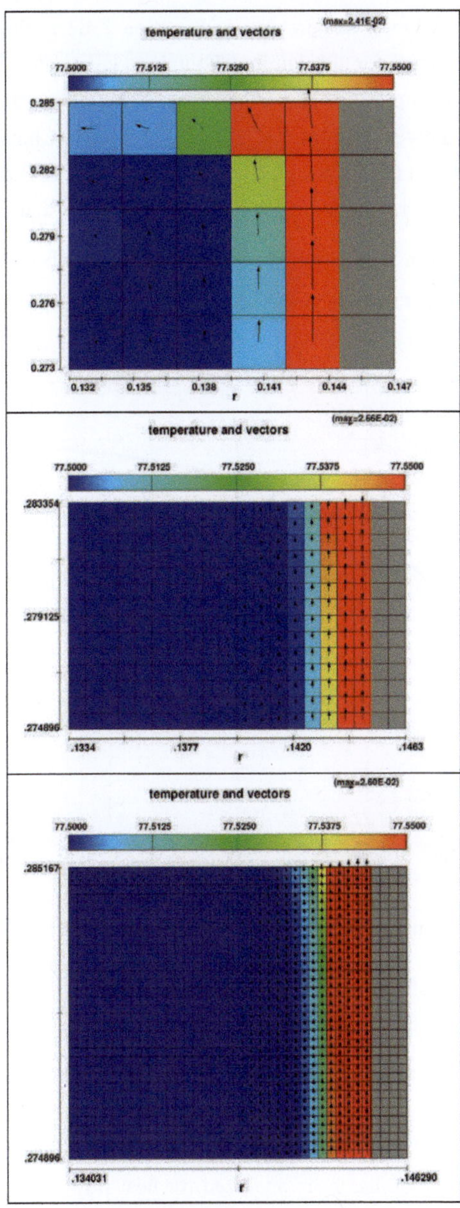

Figure 4-10. Fully developed boundary layers at t=11 s for mesh resoltions of 60x60 (top), 240x240 (middle) and 480x480 cells (bottom). Temperature scale is in K.

It is of interest to compare the development of the stratification according to FLOW 3D with the model according to Chin et al., presented in section 2.4.2.1. The results are shown in Figure 4-11. As is the case for the laminar boundary layer, an offset between the two solutions of up to 0.02 m is present. This means that FLOW 3D predicts a stratified layer of up to 0.02 m thicker at a given time than the model according to Chin et al. does.

Comparing Figure 4-11 and Figure 4-7 which each other, it is seen that in case of a turbulent boundary layer the growth rate of the stratified layer is higher than in case of the laminar boundary layer, although in both cases the heat flow $\dot{Q}_{l-n}$ has been set to the same value. This is explained by the fact that the mass flow is higher in turbulent boundary layers.

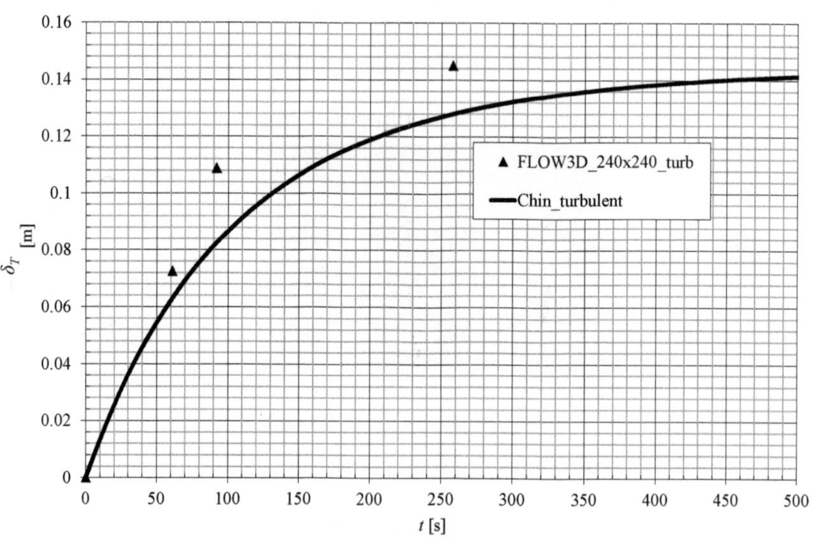

Figure 4-11. Development of stratification accoriding to FLOW3D simulations and according to Chin et al.

4.2.2 Mesh sensitivity analysis for liquid heat flow tangential through the dewar wall

Now that the mesh resolution has been analyzed for heat flow in normal direction through the dewar wall the attention must be turned to the mesh resolution required for tangential heat flow.

In section 4.1 it has been estimated that 2 W enters the liquid due to heat flow tangentially through the dewar wall. This is caused by a temperature gradient in the dewar wall in the warmer ullage region. A mesh sensitivity analysis has been performed to find out what the minimum required cell size is to obtain a solution of this process which is independent of the mesh resolution. As has been done for the mesh sensitivity analysis for heat flow normal to the dewar wall, a test case has been set up excluding the spherical bottom of the dewar. To account for the heat conduction tangentially through the wall, a part of the dewar wall (between $z = 0.29$ m and $z = 0.334$ m) in the ullage region has been included in the model. A one fluid model was used, defining only liquid nitrogen and no ullage vapor. Furthermore, no phase change was allowed. The wall temperature in this region was set to that indicated in Figure 4-1 (at $t = t_b$) as an initial condition. The wall temperature at $z = 0.334$ m was fixed at T = 115 K as a boundary condition. A schematic view of the test case is depicted in Figure 4-12.

A 2D cylindrical mesh has been used because the problem is rotationally symmetrical. Four different mesh resolutions were tried: 45x55, 120x147, 240x293 and 480x593 cells. In terms of cell size this yields 0.0032 m x 0.0032 m, 0.0012 m x 0.0012 m, 0.0006 m x 0.0006 m and 0.0003 m x 0.0003 m respectively.

Convergence is measured according to the growth rate of the thermally stratified layer δ_T, with respect to three points S1, S2 and S3. Point S1 is located at $x = 0.0725$ m and $z = 0.285$ m, S2 is located at $x = 0.0725$ m and $z = 0.27$ and S3 is located at $x = 0.0725$ m and $z = 0.25$.

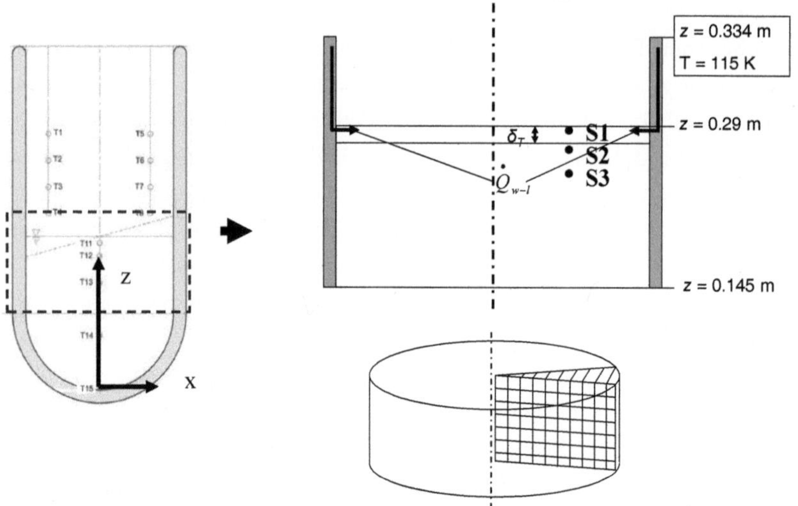

Figure 4-12. Schematic presentation of the test case used for the mesh sensitivity analysis in case of heat flow tangential through the dewar wall.

In Figure 4-13 the temperature development over 500 s for the reference points S1, S2 and S3 is given for the different mesh resolutions and laminar boundary layer settings. Convergence can be seen in the figure at a resolution of 240x293 cells, because for higher resolutions the temperature development at the reference points shows equal behavior.

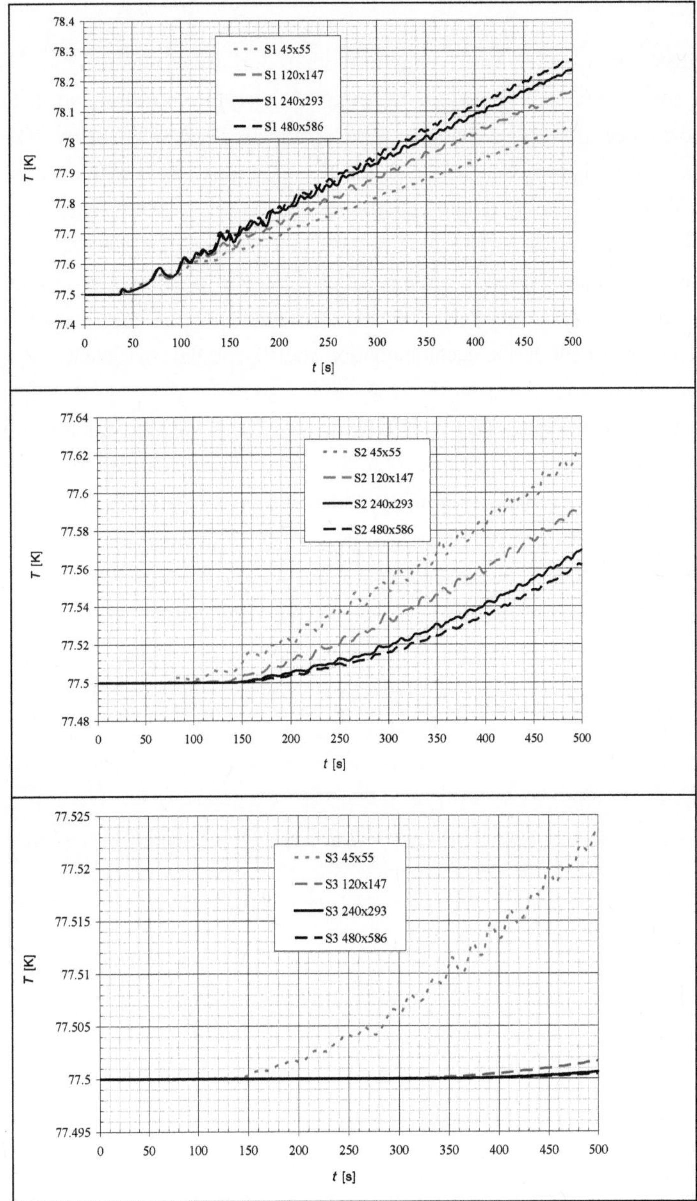

Figure 4-13. Temperature development for reference points S1 (top), S2 (middle) and S3 (bottom).

4.2.3 Sensitivity study on the accommodation coefficient

As has been shown in section 3.4.1, liquid evaporates during self pressurization. This is the main cause for the pressure rise in the system. It is therefore necessary to include phase changes in the model if pressure changes are to be modelled correctly. FLOW 3D includes a phase change model based on equation 2-49. The user must provide the accommodation coefficient. Values for accommodation coefficients have to be between 0 (no phase change) and 1 (maximum rate of phase change). Values for the accommodation coefficient are hard to determine experimentally and therefore information on accommodation coefficient is extremely hard to find in literature. To investigate the sensitivity of the pressure and temperature in the system on the accommodation coefficient, the coefficient was varied and the results are compared.

The system was modelled as a two fluid system, with the initial temperatures in the liquid set at 77.5 K and the ullage temperature set to the initial values measured during experiment sp160 (Figure 4-1). A 2D cylindrical mesh of 24x110 cells was used. This relatively low mesh resolution should suffice for the purpose of investigating the sensitivity on the accommodation coefficient. A constant heat flux of 128 W/m^2 is applied at the liquid volume, yielding at heat flow of 34 W. Just beneath the liquid surface (from $z = 0.284$ m to $z = 0.29$ m) the heat flux is increased to 4800 W/m^2 (yielding a heat flow of 26.2 W) as to let the temperature at the liquid surface rise quicker to increase the evaporation of the liquid. This is shown in Figure 4-14.

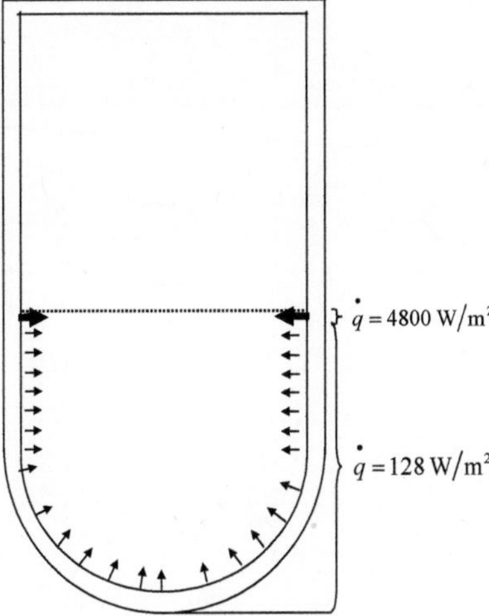

Figure 4-14. Heat fluxes defined for the sensitivity analysis on the accommodation coefficient.

Three different values for the accommodation coefficient were tried, namely 0.01, 0.5 and 0.1. Figure 4-15 shows the temperature distribution in the liquid after 400 seconds. As can be seen almost no difference can be noted. Only at the bottom of the dewar the temperature in case of the lowest accommodation coefficient shows a slight deviation. Temperatures at the liquid surface show almost no difference. The pressure rise in the system for the different accommodation coefficients is given in Figure 4-16. The pressure development shows only negligible differences, indicating that the evaporation process is largely independent of the accommodation coefficient.

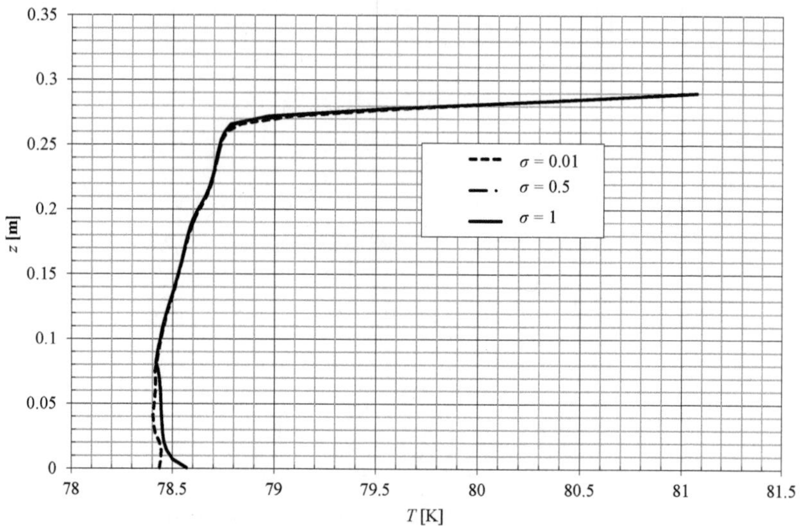

Figure 4-15. Liquid temperature distribution after 400 seconds in case of different accommodation coefficients.

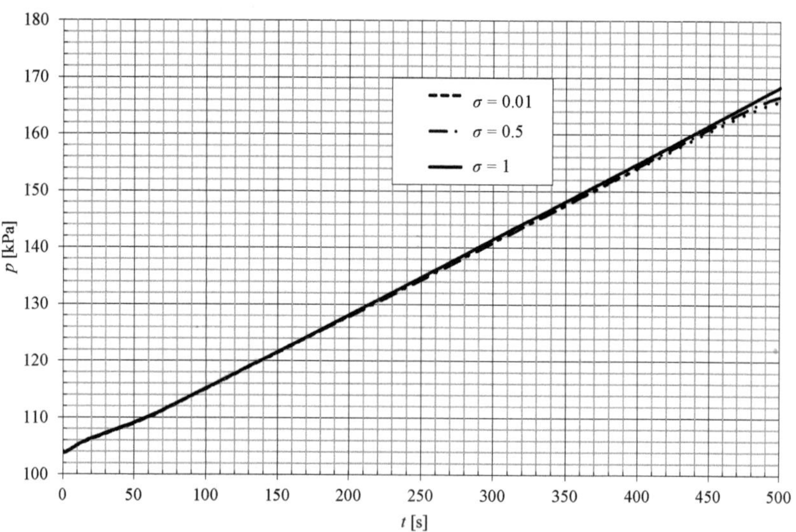

Figure 4-16. Pressure development according to different accommodation coefficients.

4.2.4 Conclusions from sensitivity analyses

Mesh sensitivity analyses were conducted for three different cases:

- Heat flow in normal direction through the dewar wall, causing a laminar thermal boundary layer to transport warm liquid to the surface. It was concluded that the thermal boundary layer thickness must be resolved by at least 5 cells, which for the presented case yields a mesh resolution of 240x240 cell or a cell size of 0.0006 m x 0.0006 m

- Heat flow in normal direction through the dewar wall, causing a turbulent thermal boundary layer to transport warm liquid to the surface. It was concluded that the thermal boundary layer thickness must be resolved by at least 5 cells, which for the presented case yields a mesh resolution of 240x240 cell or a cell size of 0.0006 m x 0.0006 m

- Heat flow in tangential direction through the dewar wall, causing the liquid at the liquid surface to heat up. It was concluded that for the presented case a mesh resolution of 240x293 cell or a cell size of 0.0006 m x 0.0006 m is required for convergence

The required cell size is 0.0006 m for all three cases. Therefore this cell size is recommended for more detailed modeling of the thermal stratification process during the experiments.

Furthermore it has been shown that the model according to Chin et al. can be used to describe the formation of thermal stratification due to heat flow in normal direction through the dewar wall.

Additionally, it has been shown that the evaporation process shows no dependency on the value of the accommodation coefficient.

4.3 Full dewar model

The full dewar model is a complex thermal model taking into account conduction through the wall (in normal and tangential direction), heat transfer into and through the liquid and ullage vapour and phase changes across the liquid vapour interface. As has been done for the test cases, the symmetry of the rotational symmetric body along the z-axis is used in the model. The use of a cylindrical mesh allows for a 2D model which is essentially a pie slice of the dewar. The opening angle of the pie slice is chosen to be 1°.

During the analysis it became clear that the recommended cell size of 0.0006 m x 0.0006 m resulted in unrealistically long computational time of several months. The smallest cell size yielding acceptable computational times was 0.003 m x 0.003 m, i.e. an increase in cell surface area of a factor 25. The consequence of this restriction is that the quantitative results must be treated with caution. One can only interpret the qualitative results by extracting for example information on heat flow mechanisms.

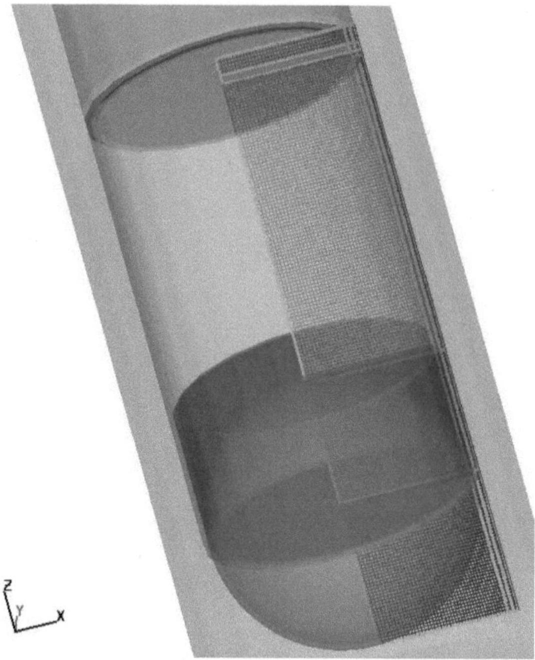

Figure 4-17. View of the dewar and the 2 dimensional cylindrical mesh.

4.3.1   Setting up the full conduction model

To simulate experiment sp160 with FLOW 3D a two fluid model was set up including a full conduction model which captures the conductive heat flow through the dewar wall. This requires creating a full model of the dewar described in section 3.1 and appendix I including the lid, the gaskets and the double walled dewar itself. The thermal conduction and heat capacity of borosilicate depend strongly on temperature, as can be seen in appendix VIII. FLOW 3D allows the user to input these temperature dependent properties in tabular form. The polyacetal lid and silicone gaskets hardly vary in temperature during the experiments, so it suffices to use room temperature properties of these materials. Dewar material properties are given in appendix VIII. The dewar is double walled, with each wall having a thickness of 6 mm. The cell size of 3 mm thus allows the thickness of each wall to be resolved by two cells.

Apart from the dewar material properties the nitrogen fluid properties specified in Table 4-1 have been used in FLOW 3D. These fluid properties are considered independent of temperature and pressure, which can be verified using appendix VII. Note that in case of gaseous nitrogen, the density is of course a function of pressure and temperature. The properties are related to each other over the ideal gas model, with $\tilde{R} = c_p - c_v$. The latent heat of evaporation $\tilde{L}$ shows limited but noticeable variation with respect to pressure. Therefore the average value of the latent heat at 100 kPa (199.3E03 J kg^{-1}) and at 160 kPa (193.7E03 J kg^{-1}) is used.

As has been discussed in section 4.2.3, FLOW 3D uses equation 2-49 to model the phase change. The accommodation coefficient has been set to 0.01. The saturation curve defined in FLOW 3D is presented in section 2.3.1.

| | ρ | c_v | c_p | β | μ | λ | $\tilde{L}$ |
	[kg m^{-3}]	[J kg^{-1} K^{-1}]	[J kg^{-1} K^{-1}]	[K^{-1}]	[Pa s^{-1}]	[W m^{-1} K^{-1}]	[J kg^{-1}]
Liquid	806	1083	2042	0.0057	160E-6	0.15	196.5E03
vapour	NA	7050	1050	0.0064	10.7E-6	0.02	196.5E03

Table 4-1. Nitrogen fluid properties used in FLOW 3D.

The initial temperature in the liquid is approximated as constant in space and equal to 77.5 K. The initial ullage temperature has been set according to Figure 4-1.

The inner and outer walls are connected with each other at the top of the dewar. The outer wall is assumed to remain constant at room temperature. The initial temperature of the inner wall is assumed to be equal to the fluid temperature. Heat is thus conducted from the hotter outer wall via the connection at the top to the inner wall. The silverized vacuum space between the walls is hard to model exactly. The vacuum will prevent conductive heat transport between the inner and outer wall and the silver will minimize heat transported by radiation, but zero heat transfer will never be reached. The manufacturer of the dewar (KGW Isotherm) has been contacted but couldn't provide any details about the heat transfer over the vacuum space. It is possible however, to determine this indirectly from experimental data. In section 3.4.1 the flow in to the liquid has been determined at $\dot{Q}_l = 6.7$ W. In the previous section the heat flow from the ullage region into the liquid has been estimated at $\dot{Q}_{ul-l} = 2.5$ W. This leaves a heat flow of $\dot{Q}_{l-n} = \dot{Q}_l - \dot{Q}_{ul-l} = 4.2$ W which travels normal to the dewar wall through the vacuum space and enters the liquid. Assuming that the outer wall has a temperature of 290 K and the inner wall has a temperature of 77.5 K in the liquid region the overall heat transfer coefficient over the vacuum can be determined as $h_{vac} = \dfrac{\dot{Q}_{l-n}}{S_{w-l}(T_{w-o} - T_{w-i})}$, where S_{w-l} is the wall surface area in contact with the liquid (0.264 m^2), T_{w-o} is the outer wall temperature and T_{w-i} is the temperature of the inner wall in the liquid region. Inserting the values yields $h_{vac} = 0.072$ Wm^{-2}K^{-1}. Flow 3D gives the user the possibility to define heat transfer coefficient between components so h_{vac} was set as the heat transfer coefficient between outer and inner wall in FLOW 3D.

Figure 4-18 shows the FLOW 3D component model of the dewar. As can be seen the mesh size allows the inner and outer parts of the wall to be resolved by two cells. The left part shows a schematic view of the region under consideration in this figure. It is the same as the highlighted region in appendix I. The middle part indicates the different components within the model. In the model the inner and outer wall are not separated by a vacuum space. This is not necessary the borosilicate arc connecting the inner and outer wall. In the right part of the figure the initial temperatures of the components because the heat transfer between the two walls is defined by h_{vac}. The "connection" is actually just the borosilicate arc connecting the inner and the outer wall. In the right part of the figure the initial temperatures of the components can be seen. The lid is given a fixed temperature of 285 K, as is obtained from experimental data. The gasket is given an initial temperature of 285 K, but is allowed to vary in time (by specifying the specific heat, density and conduction coefficient). The outer wall has a fixed temperature of 290 K and the inner wall has an initial temperature distribution equal to that of the initial fluid temperature (see Figure 4-1) but is allowed to vary in time.

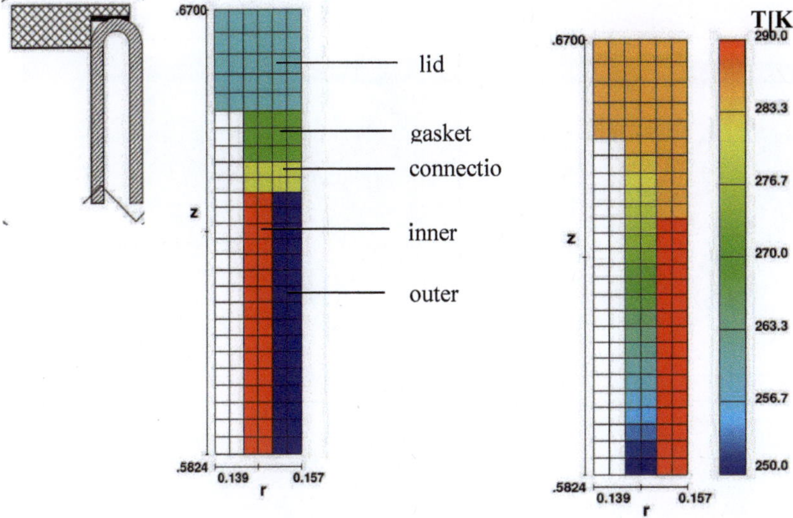

Figure 4-18. Numerical model of the dewar.

The different components in the numerical model are indicated in the middle part, initial temperatures of the components in the right.

4.3.2   Results of the full conduction model

The numerical results will be discussed focussing on the liquid. The initial temperature in the liquid is approximated as constant in space and equal to 77.5 K. In Figure 4-19 the temperature in the liquid during the stratification process is visible. It can clearly be seen that due to heat entering the liquid via the dewar wall, a thermal boundary layer at the wall develops. The heated liquid in this boundary layer then rises to the liquid surface, creating a pocket of warm liquid at the liquid surface. The pocket volume increases over time and after 300 s the complete liquid region is heated. At the bottom of the dewar a column of warm liquid builds up, also transporting warmer liquid to the liquid surface. Over time a strong thermal gradient at the liquid surface develops. After 3725 s (the end of the pressurization phase of sp160) the thickness of the thermal gradient, as well as the temperature at the liquid surface have reached a maximum. Temperature at the liquid surface has reached 80.63 K, bulk liquid temperature lies at about 78.1 K. What is also interesting to see is that over time the liquid temperature shows no variation over the radial direction anymore.

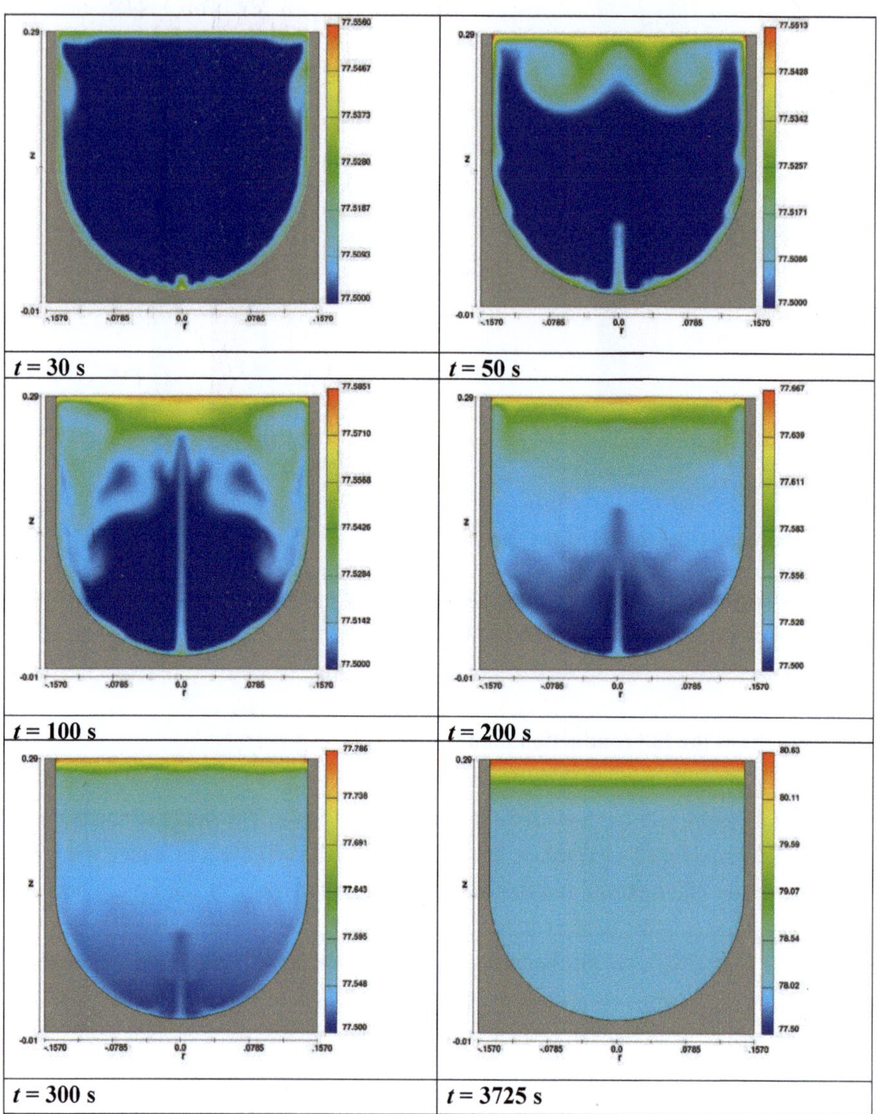

Figure 4-19. Numerical simulation of development of thermal stratification in the liquid. Only the liquid region of the dewar is shown. The temperature scale is in K.

The heat flux entering each liquid cell in contact with the wall can be seen in Figure 4-20. The right hand side of Figure 4-20 shows the enlargement of the grid near the wall at the liquid surface. It is clearly seen that just below the liquid surface the cell heat flux is much higher than at the rest of the wall. This is caused by the aforementioned heat conduction in tangential direction through the dewar wall. This heat mainly enters the liquid at the top 5 cells, as can be seen in the right part of Figure 4-20. From these cell heat fluxes and cell dimensions the heat flow entering the liquid in the top 5 cells can be determined, resulting in $\dot{Q}_{w-l} = 2.7$ W. This is 0.7 W more than was estimated in section 4.1.

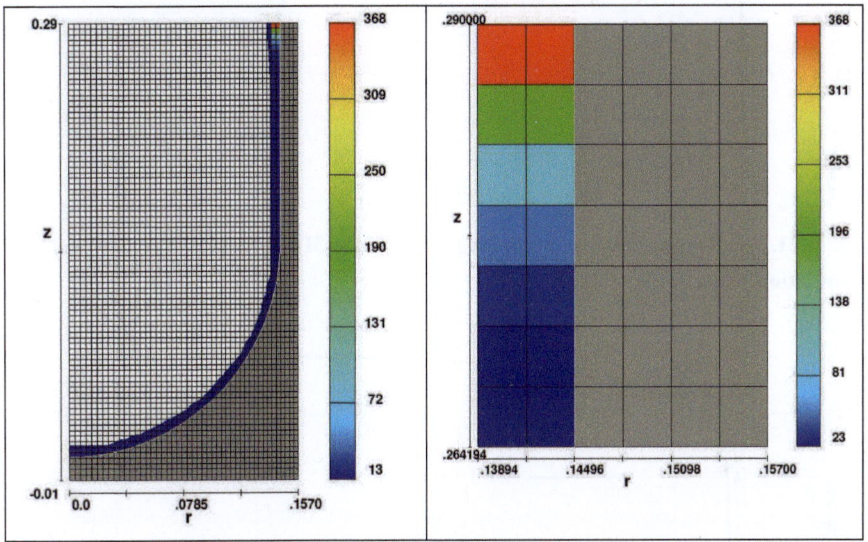

Figure 4-20. Heat fluxes into the liquid. The right hand side show an enlargement of the top right hand corner of the dewar. The temperature scale is in K.

The temperature distribution in the dewar wall causing this conductive heat flow is shown in Figure 4-21. The wall temperature distribution remains approximately the same, i.e. an approximately linear distribution between lid and liquid surface. During the pressurization period, the magnitude of the temperature increases by about 6 K in the ullage region. Figure 4-22 gives the numerical temperatures at the end of the pressurization phase for the wall in the liquid region and the temperatures of the liquid itself. The wall temperature near the liquid surface shows a temperature difference of approximately 4K with the liquid temperature.

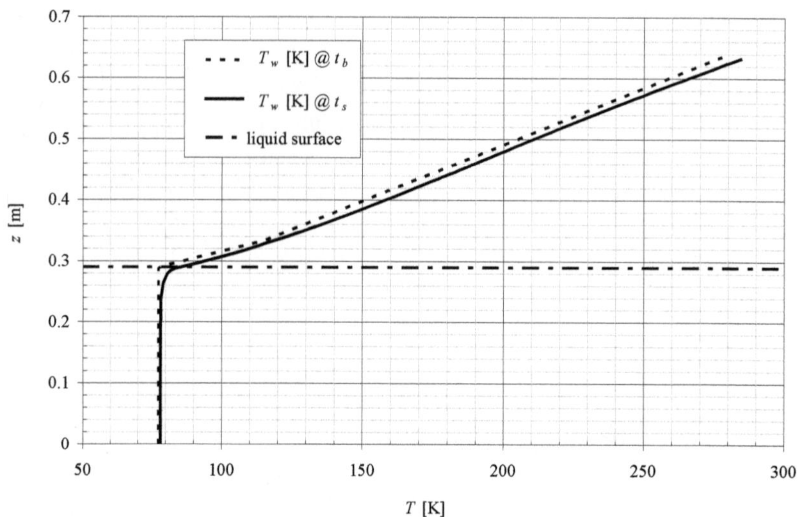

Figure 4-21. Wall temperature according to FLOW 3D at beginning and end of pressurisation phase for sp 160.

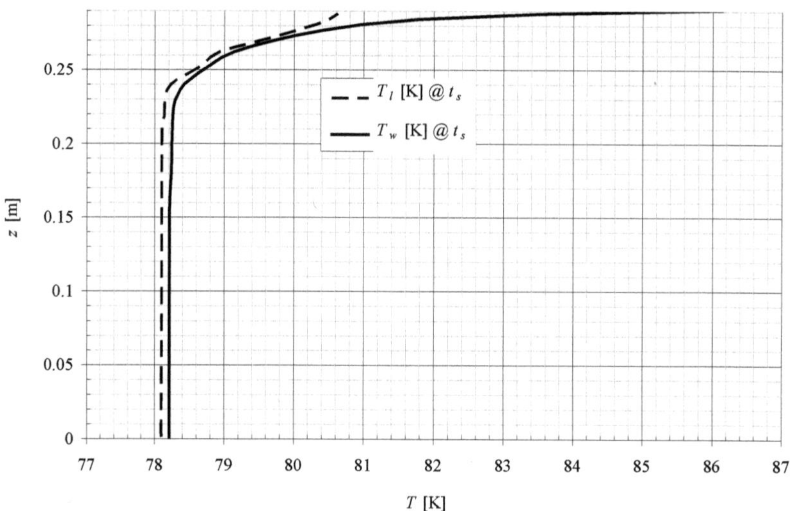

Figure 4-22. Numerical wall temperatures at the end of the pressurization phase vs. numerical liquid temperature.

The temperature distribution in the liquid at the centerline of the dewar ($x = 0$ m) is analyzed in more detail using Figure 4-23. At $t = 100$ s it is seen that the near the liquid surface the liquid temperature has increased. This is mainly caused by the warmer liquid transported upward in the thermal boundary as a result of $\dot{Q}_{l-n}$. Near the bottom of the dewar ($z = 0$ m), the liquid is also heated, which is caused by the column of heated liquid also visible in Figure 4-19. At $t = 300$ s the formation of a strong thermal gradient near the liquid surface can be seen, which is caused by $\dot{Q}_{w-l}$. The thickness of this thermal layer increases over time. At $t = 300$ s it is estimated at $\delta_T = 0.03$ m and at $t = 3725$ s it is estimated at $\delta_T = 0.06$ m. The thickness increases by conductive heat flow through this thermal gradient into the liquid bulk (thermal diffusion).

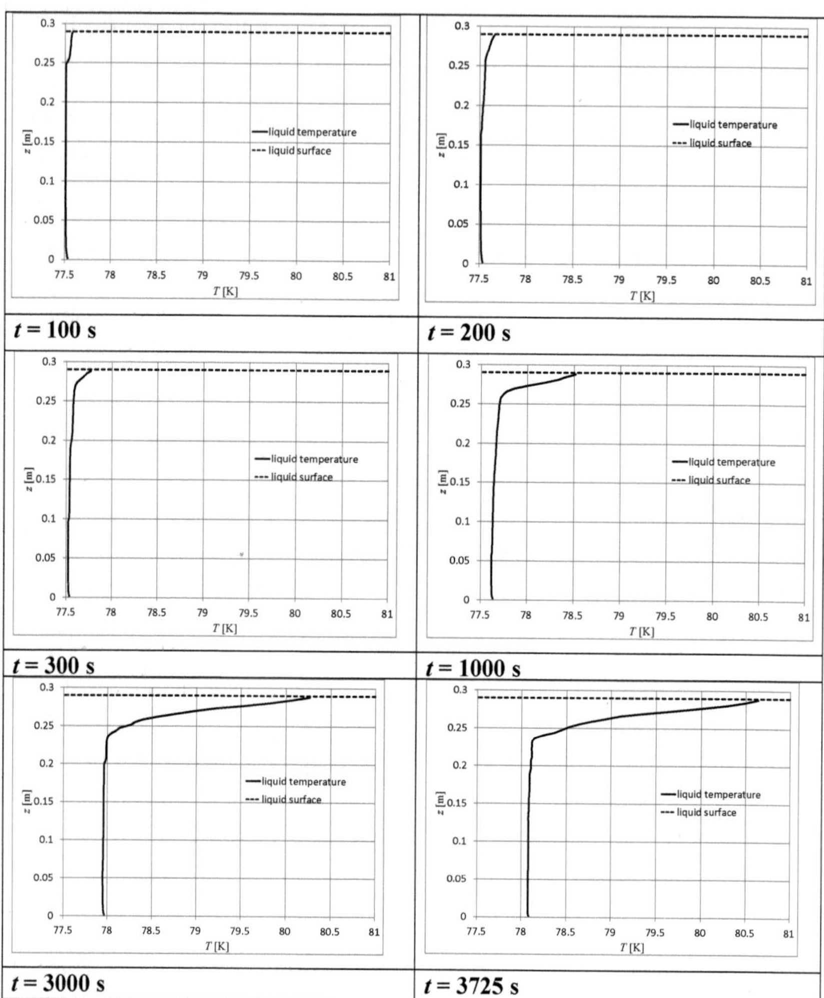

Figure 4-23. Numerical simulation of temperature development in the liquid.

Numerically obtained temperatures at the end of the pressurization phase of sp160 are compared to the temperatures measured during the experiment in Figure 4-24. The thermal gradient between $z = 0.253$ m and $z = 0.278$ m is captured very well, but at the liquid surface there is a temperature difference between numerics and experiment of 0.8 K. The most likely reason for this is that the mesh resolution is not high enough to yield reliable temperature data near the liquid surface.

For the sake of completion the temperature distribution in the complete dewar at end of the pressurisation phase is shown in Figure 4-25. The pressure development is shown in Figure 4-26. Because of the saturation conditions at the liquid surface the lower pressure for the numerical results agrees with the lower temperature at the liquid surface.

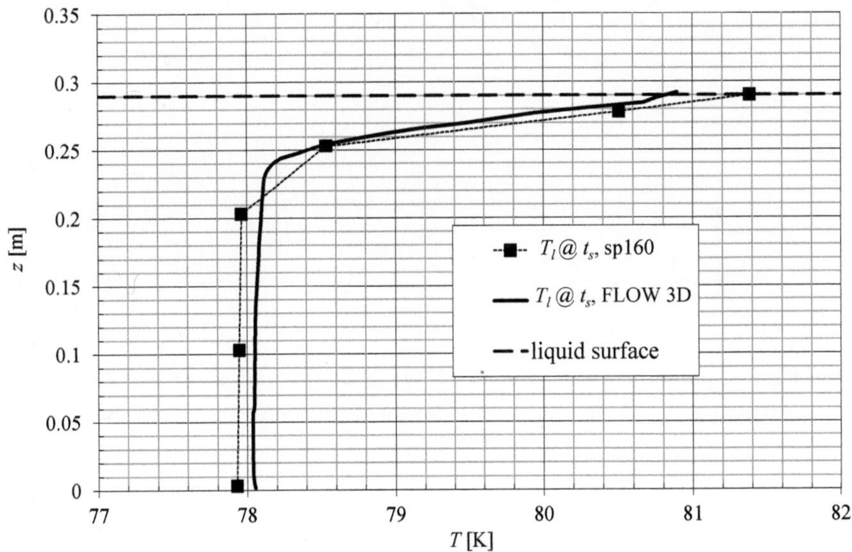

Figure 4-24. Temperature distribution in the liquid at the end of the pressurization of experiment sp160 compared to the numerical results.

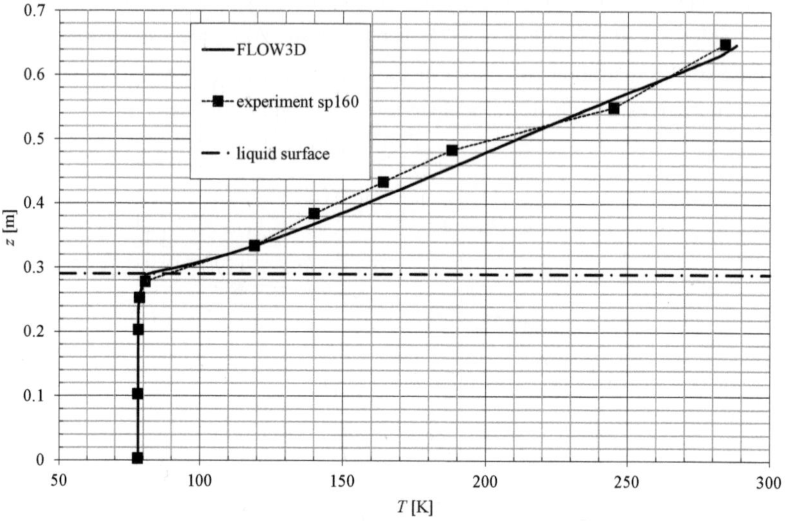

Figure 4-25. Temperature distribution in the complete dewar at the end of the pressurization phase of sp160 compared to the numerical results.

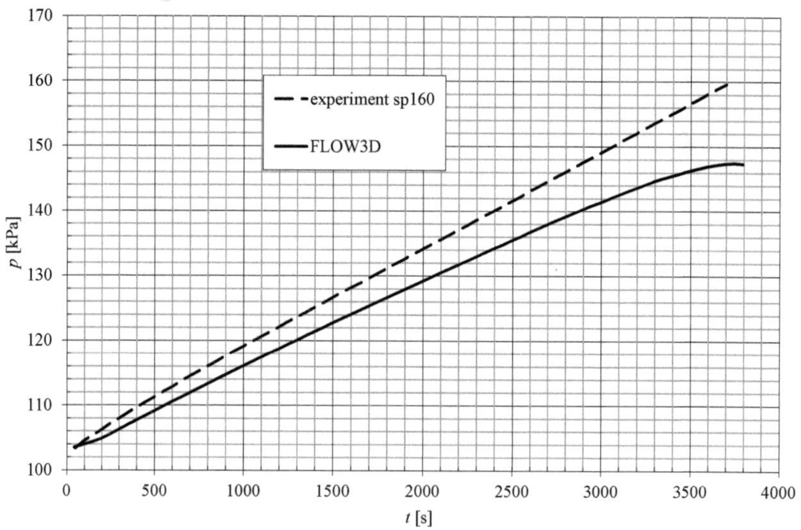

Figure 4-26. Pressure development during experiment sp160 compared to the numerical results.

4.3.3 Conclusions from the FLOW3D full conduction model

Although the required mesh resolution for the FLOW 3D model could not be implemented due to unacceptably long computational times, the results have yielded some important qualitative conclusions with respect to the formation of thermal stratification. These conclusions are summarized below:

- The temperature in the radial direction is approximately homogeneous after already 300s

- The heat flow normal through the dewar wall ($\dot{Q}_{l-n}$) causes a thermal boundary to transport heated liquid up towards the liquid surface.

- The volume of this heated pocket of liquid increases over time. The temperature of this pocket is approximately homogeneous. As soon as the volume of the pocket is equal to the liquid volume (i.e. the complete liquid volume has been heated), the temperature thus has reached an approximately homogenous condition (bulk temperature).

- The heat flow tangentially through the dewar wall ($\dot{Q}_{w-l}$) causes heating of the liquid just under the liquid surface. This mechanism is responsible for the sharp thermal gradient near the liquid surface.

- The thickness of this thermal gradient increases slowly over time due to heat conducted through this thermal layer into the bulk of the liquid.

4.4   A 1D engineering model for the simulation of thermal stratification in the liquid

The full 3D model could not be used to its full potential simply because the required mesh detail would results in unrealistic computational times. It is therefore of interest to see if a simpler 1D model could generate similar results without suffering from the practical restrictions in computation time.

This section describes a 1D engineering model, developed to simulate the stratification process of the experiments very fast and without the need of a CFD program. For accurate modeling of the stratification, it is necessary to solve for the thermal conduction within the fluid and the wall. Therefore, a 1D conduction model has been developed. In the paragraphs below the 1D model is described followed by a discussion of the results obtained from the simulation of experiment sp160. Finally the 1D model is applied to the active pressurisation experiment ap160 and the results are discussed.

4.4.1   Thermal conduction model

One of the conclusions in section 4.3.3 is that the liquid temperature reaches a homogeneous distribution in the radial direction. This implies that an engineering model does not need to solve for temperature variations in the radial direction. It is also known that the thickness of the thermal gradient near the liquid surface increases over time due to conduction into the liquid bulk. This is described by 1D heat transfer based on equation 2-21:

$$\rho_l c_{p,l} \frac{\partial T_l}{\partial t} = \lambda_l \frac{\partial^2 T_l}{\partial z^2} \qquad\qquad 4\text{-}1$$

where $\mathbf{u}$ has been set to zero. The boundary condition at the liquid surface is described by equation 2-22 (with $\mathbf{u}_l = 0$) together with the assumption of saturation conditions at the liquid surface ($T_l = T_v = T_{sat}(p_v)$).

The boundary condition at the wall is described by equation 2-32, where it reduces to $\overset{\bullet}{\mathbf{q}} = -\lambda \dfrac{\partial T}{\partial \mathbf{n}}\bigg|_w$ and can conveniently be determined using equation 2-34.

The energy equation can be solved numerically by dividing the liquid region in a number of nodes (layers) in the z- direction. Because it has been shown in section 4.1 that tangential heat transport through the wall is a major contributor to the thermal stratification, it is necessary to solve the energy equation for the wall as well. The energy equation for the wall is just equation 4-1 but with wall material properties instead of liquid properties, i.e. the subscript "l" must be replaced by "w". In this way the boundary condition at the wall can be solved using equation 2-34 and the liquid and the wall temperatures, yielding the heat flow into the liquid.

Because the model is divided in two nodes in the x direction (one for the liquid and one for the wall), the model is strictly speaking a 2D model. But because there are only two nodes in the x direction and there are hundreds of nodes in the z direction the model is referred to as a 1D model.

Details are shown in Figure 4-27, the liquid region has been divided in n layers. The gray nodes on the right side represent the wall and the nodes on the left side represent the liquid. To simplify the model, only that part of the system below the liquid surface is taken into consideration. This means that all heat coming from above the liquid surface is a boundary condition for the model.

Each node has a non-constant temperature, but wall and fluid properties are assumed constant. The wall properties are assumed to be equal and constant for all nodes. Also the liquid the liquid properties are assumed equal and constant for all nodes. This approximation is valid because the variation of temperatures below this height is not high enough to influence the material and fluid properties significantly.

In Figure 4-27 conductive heat flow between the nodes is indicated by a black solid arrow. The arrows indicate the direction of positive heat flow. The open arrows indicate external heat flow acting on the system as boundary conditions. In this case these heat flows contain the experimentally determined heat flow normal through the wall ($\dot{Q}_{l-n}$). For the node located at the liquid surface, this also includes the heat transported through the vapor into the liquid ($\dot{Q}_{v-l}$). For the wall node located at this height, this includes the heat conducted tangentially through the wall ($\dot{Q}_{w-l}$).

The boundary condition at the liquid surface, i.e. equation 2-22 together with the saturation conditions at the liquid surface, is taken into account using equation 2-49 with $\sigma = 0.01$ and where $p_{sat,l}$ is calculated using equation 2-42.

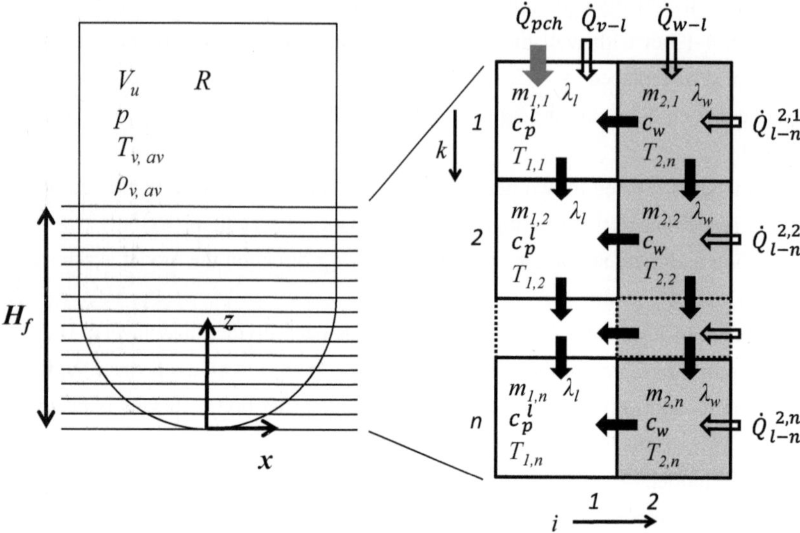

Figure 4-27. Heat conduction model.

As an initial condition, the wall temperature is assumed to be equal to that of the initial fluid temperature. The boundary conditions will cause heating of the nodes and the conductive heat flow will start.

Equation 4-1 can be modelled in a discreet way using the following heat conduction equation [23]:

$$\dot{Q}_{cond}^{i,k} = S \frac{T_{i,k} - T_{i,k+1}}{d} \lambda \text{ for heat flow in } z \text{ direction, } \lambda = \lambda_l \text{ for } i = 1 \text{ and } \lambda =$$

$$\lambda_w \text{ for } i = 2 \qquad\qquad 4\text{-}2$$

$$\dot{Q}_{cond}^{k} = S \frac{T_{2,k} - T_{1,k}}{d} \lambda_l \text{ for heat flow in } x \text{ direction}^2$$

---

[2] Here λ_l is taken instead of λ_w because λ_l is smaller, so it is the limiting factor in the heat transfer. Furthermore, it is noted that determining the heat flow in x direction using equation 4-2 may not be very accurate as there are only two nodes in this direction. A more accurate approach would be to use $\dot{Q}_{cond}^{k} = hS(T_{2,k} - T_{1,k})$, where h must be determined experimentally.

where d is the distance between the centre points of the nodes and S is the contact surface area between the nodes. In the model all nodes are assumed to have the same size, so these parameters are constant. The temperature increase in a node can be calculated by:

$$T_{2,k}^{new} = T_{2,k} + \frac{\dot{Q}_{cond}^{2,k-1} - \dot{Q}_{cond}^{2,k} - \dot{Q}_{cond}^{k} + \dot{Q}_{ex}^{2,k}}{m_{2,k}c_p}\Delta t \quad \text{and} \quad c_p = c_w \text{ for } i = 2$$

$$T_{1,k}^{new} = T_{1,k} + \frac{\dot{Q}_{cond}^{1,k-1} - \dot{Q}_{cond}^{1,k} + \dot{Q}_{cond}^{k} + \dot{Q}_{ex}^{1,k}}{m_{2,k}c_p}\Delta t \quad \text{and} \quad c_p = c_{p,l} \text{ for } i = 1$$

4-3

At node (1,1) $\dot{Q}_{ex}^{1,1} = \dot{Q}_{v-l} + \dot{Q}_{pch}$. At node (2,1) $\dot{Q}_{ex}^{2,1} = \dot{Q}_{w-l} + \dot{Q}_{l-n}^{2,1}$. At nodes $(2,n\ \ n \neq 1)$ $\dot{Q}_{ex}^{i,k} = \dot{Q}_{l-n}^{2,n}$. For all other nodes $\dot{Q}_{ex}^{i,k} = 0$. Heat transfer normal to the wall $\dot{Q}_{l-n}^{2,n}$ is determined as $\frac{\dot{Q}_{l-n}}{n}$. It causes the formation of a thermal boundary layer which can be modeled according to equation 2-66 and 2-67. For the experiments at ZARM a turbulent boundary layer is expected (see section 4.1) so equation 2-66 should be used. The temperature distribution in the heated volume is determined using equation 2-68 with $I = 0.5$. Based on the results of the numerical simulations, it is assumed that as soon as $\delta_T = H_l$, the liquid heats up uniformly.

Modeling of the ullage pressure

The ullage is modelled using only a single node with a certain initial pressure and temperature. Pressure change in the ullage is modelled using equation 3-2. The rate of density change is determined as:

$$\frac{d\rho}{dt} = \frac{\dot{m}_{pch} + \dot{m}_i}{V_u}$$

4-4

The phase change mass flow rate $\dot{m}_{pch}$ is determined by equation 2-49 and coefficients defined in section 2.3.1, where $p_{sat,l}$ is the saturation pressure of the liquid temperature in the node at the liquid surface (node 1,1). The saturation pressure is determined using equation 2-42. The ingoing mass flow rate $\dot{m}_i$ is the mass added actively to the ullage, and is thus zero in case of self pressurization.

The rate of temperature change can be determined by setting up an energy balance for the ullage based on equation 2-14 (with $\dot{W}_{cv} = 0$) yielding:

$$\dot{I}_v = m_v c_v \, \dot{T}_{v,av} + \dot{m}_v \, c_v \, T_{v,av} = \dot{Q}_{cv} + \dot{m}_{pch} \, h_e + \dot{m}_i \, c_p T_i \quad \text{[3,4,5]} \qquad \text{4-5}$$

The heat flow rate $\dot{Q}_{cv}$ is the heat flow rate into the ullage $\dot{Q}_v$ which has been determined section 3.4.2 to be 0.7 W. Rewriting this equation results in:

$$\dot{T}_{v,av} = \frac{\dot{Q}_v + \dot{m}_{pch} \, h_e + \dot{m}_i \, c_p T_i - \dot{m}_v \, c_v \, T_{v,av}}{m_v c_v} \qquad \text{4-6}$$

Determining the heat flow from the ullage into the liquid
The boundary condition at the liquid surface is described by equation 2-22:

$$\dot{Q}_{pch} = S_{ls} \tilde{L} \rho_l (u_{z,l} - u_{z,\Gamma}) = \tilde{L} \, \dot{m}_{pch} = S_{ls} \lambda_v \left. \frac{\partial T_v}{\partial z} \right|_{ls} - S_{ls} \lambda_l \left. \frac{\partial T_l}{\partial z} \right|_{ls} \qquad \text{4-7}$$

It is convenient to write equation 4-7 as $S_{ls} \lambda_l \left. \dfrac{\partial T_l}{\partial z} \right|_{ls} = S_{ls} \lambda_v \left. \dfrac{\partial T_v}{\partial z} \right|_{ls} - \dot{Q}_{pch}$, where the 2nd and 3rd term represent heat extracted from the ullage. The first term on the right hand side is just $\dot{Q}_{v-l}$ which has been determined at 0.5 W in section 4.1.

---

[3] The approximation $I_v = m_v c_v T$ has been used.

[4] This equation can also be written as $\dot{m}_{pch}(c_v T_{ls} - h_e) + \dot{m}_{pch} \, c_v (T_{v,av} - T_{ls}) + \dot{m}_v \, c_v \, \dot{T}_{v,av}$, where the first term corresponds to the latent heat.

[5] Note that $\dot{m}_{pch}$ has a positive sign even though the mass is exiting the system. This is because condensating mass already has a negative value according to equation 2-49.

The heat extracted due to phase changes is determined using equation 4-7; $\dot{Q}_{pch} = \tilde{L}\,\dot{m}_{pch}$.

4.4.2 Results of the 1D model applied to the experiments

The model described above is applied to experiments sp160 and ap160. Before the results are discussed the required input and the initial conditions are given.

- Fluid property data.

  The fluid data used in the model is listed in Table 4-1. The gas constant $\tilde{R} = c_p - c_v$. For modeling of phase changes the saturation properties as presented in section 2.3.1 are used. The accommodation coefficient is set to $\sigma = 0.01$.

- Wall material properties.

  Wall material properties are taken from appendix VIII at $T = 77.5$ K and are $\lambda = 0.45$ W m^{-1} K^{-1}, $c = 200$ J kg^{-1} K^{-1}, $\rho = 2200$ kg m^{-3}. As only that part of the wall which is in contact with the liquid is considered in the model, the material properties of the wall can be assumed constant (the temperature in this region will not vary enough to have significant influence on the material properties).

- Initial conditions.

  Initial pressure of the system is assumed at 1.03 bar. The initial temperature of the ullage is the average vapour temperature $(T_{v,av})$ at the beginning of the pressurization phase presented in Figure 3-6 and Figure 3-11. Initial liquid temperature is assumed homogeneous at 77.5 K. The initial wall temperature is assumed equal to the liquid temperature.

- External heat fluxes.

  In section 4.1 it is argued that $\dot{Q}_{l-n} = 4.2$ W, $\dot{Q}_{w-l} = 2$ W and $\dot{Q}_{v-l} = 0.5$ W.

- Geometrical data.

  The ullage volume $V_u = 0.0228 \pm 1.3\text{E-4 m}^3$ where V_u is determined from geometrical data in Figure 3-1. The liquid fill height is taken from section 3.1, $H_l = 0.29$ m.

- Number of layers.

  The liquid has been divided into $n = 1200$ layers. The distance $d = H_l/1200 = 2.42\text{E-4}$ m.

Self pressurization

Experiment sp160 is simulated using the 1D model and the input defined above. Results of the 1D model are compared to experimental results and FLOW 3D results in Figure 4-28, Figure 4-29 and Figure 4-30. Figure 4-28 shows the pressure development according the FLOW 3D, the 1D model and as measured during the experiment. The results of the 1D model are in closer agreement to the experiment than the FLOW 3D results are. FLOW 3D underestimates the pressure, which could be caused by the suboptimal grid size (see section 4.3).

Figure 4-29 shows a comparison of the liquid temperature. The thermal gradient at the liquid surface is captured well by both models. However, the temperature at the liquid surface is about 0.5 K lower for the FLOW 3D model, which agrees with the lower pressure. The 1D model results show minimal deviation compared to the experimental results.

Figure 4-30 compares the wall temperatures in the liquid region calculated by the 1D model and FLOW 3D. The wall temperature according to FLOW 3D is about 0.2 K higher than the temperatures of the 1D model.

CPU time for the 1D model is 99 seconds, whereas the CPU time for the FLOW 3D model is 44 hours.

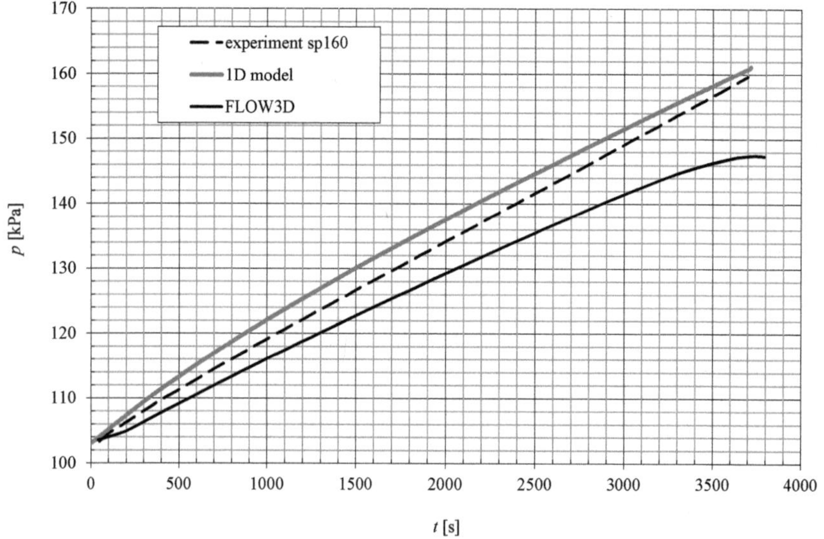

Figure 4-28. Pressure development according to experiment sp160, FLOW 3D and 1D model.

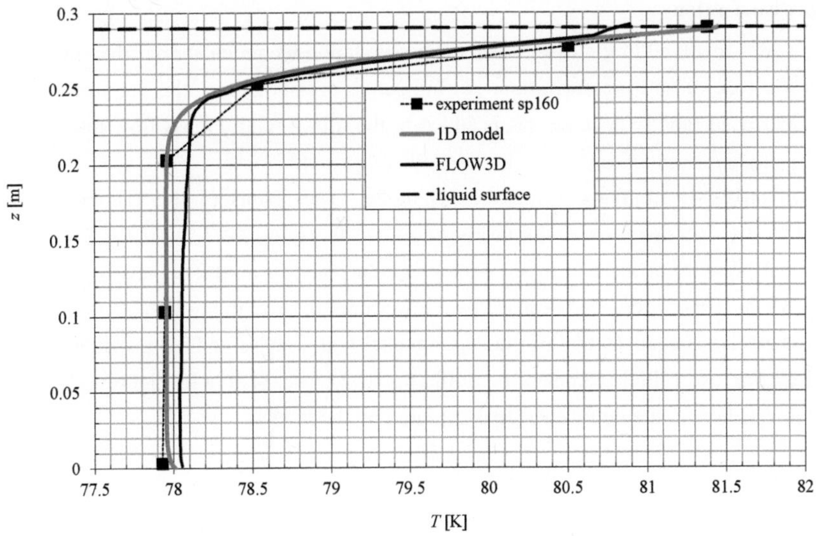

Figure 4-29. Temperature in the liquid at the end of the pressurisation phase according to experiment sp160, FLOW 3D and 1D model.

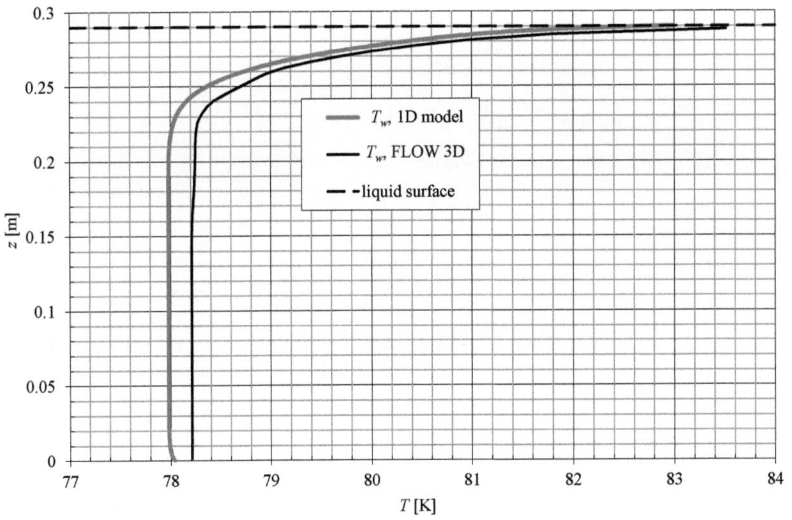

Figure 4-30. Wall temperature at end of pressurisation phase for sp160, according to FLOW 3D and 1D model.

Active pressurization

The time scales involved in the case of active pressurization are much shorter than the time scales in the case of self-pressurization. During active pressurization, the pressure in the ullage is increased rapidly because gas is added to the ullage with a mass flow rate of 0.2 g/s.

Experiment ap160 is simulated using the 1D model. The results of the 1D model are compared to the experimental results in Figure 4-31 and Figure 4-32. Figure 4-31 shows that the final pressure at the end of the pressurization phase is underestimated by the 1D model by 8 kPa. This difference is caused by a lower pressure rise in the first 20 s. After 20 s, the pressure rise is equal to that of the experimental one. The lower final pressure is also reflected by the lower liquid temperature at the liquid surface, visible in Figure 4-32. The temperature gradient at the liquid surface is captured by the model.

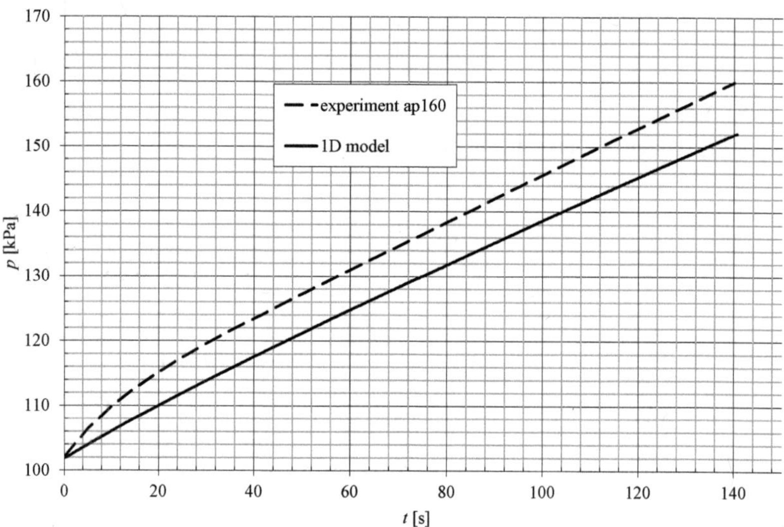

Figure 4-31. Pressure development according to experiment ap160 and the 1D model.

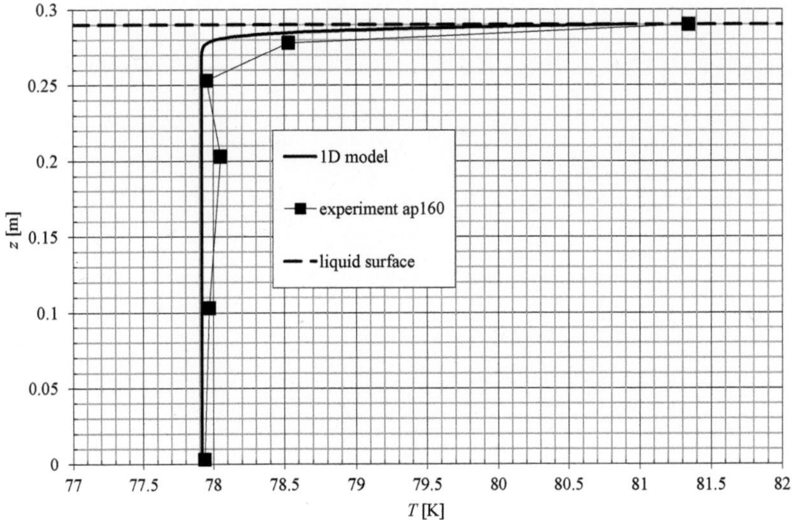

Figure 4-32. Temperature in the liquid at the end of the pressurisation phase according to experiment ap160 and the 1D model.

5 Numerical modeling of sloshing including heat and mass transfer

Now that the thermal stratification, the main cause of pressure drop during sloshing, has been numerically simulated the attention can be turned to the numerical simulation of the pressure drop due to sloshing.

In this chapter the sloshing simulation with FLOW 3D is discussed first (section 5.1). It will be shown that it is not possible to obtain reliable results in FLOW 3D due to limitations in the numerical methods used in FLOW 3D. The reasons behind this failure are explained in detail. Next, the 1D model, developed in the chapter 4, will be extended to enable pressure drop during sloshing. The simulation results of this 1D model, which are considerably better than the FLOW 3D results, are discussed in section 5.3.

5.1 Numerical modeling in FLOW 3D

FLOW 3D simulations were started at the beginning of the sloshing ($t = t_s$). Initial conditions of the simulations, such as temperature distribution and pressure, were taken from the experiments (see chapter 3). The experiments were modelled as an adiabatic two fluid system, i.e. no heat enters or exits the tank. Experimental settings were chosen such that a well-defined planar, non-rotating slosh motion was present. This allows simplifying the numerical modeling by taking into account symmetry along the xz plane. But in contrary to the simulation of thermal stratification discussed in chapter 4, the problem is not rotationally symmetric so a 3D mesh is required.

Three different mesh resolutions were tried to investigate mesh dependency. The first and lowest mesh resolution consisted of 29250 cells (30x15x65). Cell number was increased in the second mesh by a factor 3.5 yielding a 101430 cell mesh. Finally, the third and highest mesh resolution was obtained by again multiplying the total number of cells by 3.5, yielding a total of 351000 cells. Slosh motion was introduced by applying acceleration in the x –axis direction by tabular input. The tabular values were taken from acceleration values measured during the experiments.

Figure 5-1 shows the results of these simulations. All cases give unsatisfactory results in the simulation of the pressure development. The course 29250 cell mesh yields the biggest error in pressure development. The mesh with a higher resolution (101430 cells) yields a slight improvement. By again increasing the number of cells 3.5 times (351000 cells) the improvement is much smaller indicating that the result has converged, even though the simulated pressure drop is to large compared to the experimental results.

To investigate what causes the pressure drop in the numerical simulations, the phase change model was switched off. By switching off the phase change model the average density in the ullage must remain constant. This excludes the possibility for a pressure change due to changes in the ullage density. In addition, the heat conduction coefficients of the liquid and vapour were set to zero and no heat exchange was allowed between the liquid and vapour. Because there is no heat transfer between liquid and vapour and the system is modelled adiabatically, the average temperature in the vapour should remain constant. A pressure drop due to temperature decrease in the ullage is therefore also ruled out. Pressure in the system should remain constant during sloshing.

The result has also been plotted in Figure 5-1 and it can be seen that with these settings, a considerable pressure drop is still present. By taking a look at Figure 5-2, where numerically predicted ullage temperature is plotted, it can be seen that temperature in the ullage according to the numerical model drops, although no heat transfer was allowed. This numerical cooling of the ullage vapour causes the unrealistic pressure drop in the system. The cause for this temperature drop was found to be a numerical error introduced by the numerical methods used in the software. In the next section this will be explained in detail by using a simplified version of the FLOW 3D model.

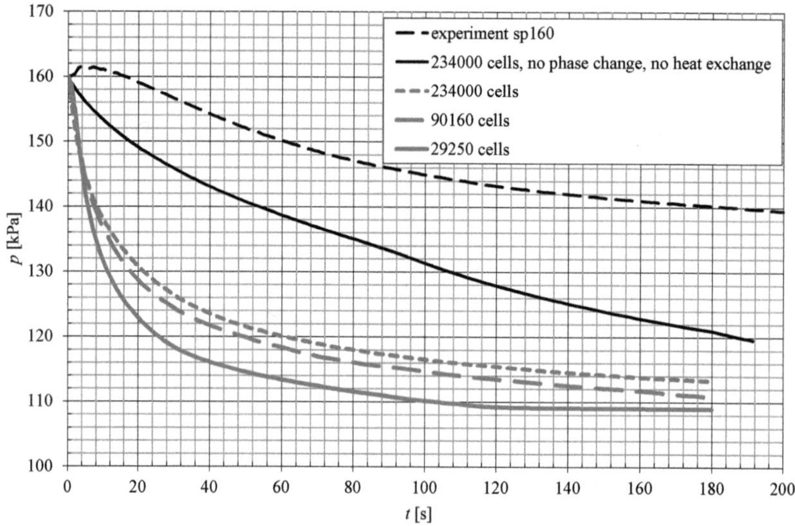

Figure 5-1. Numerical pressure development compared to the experimental pressure development.

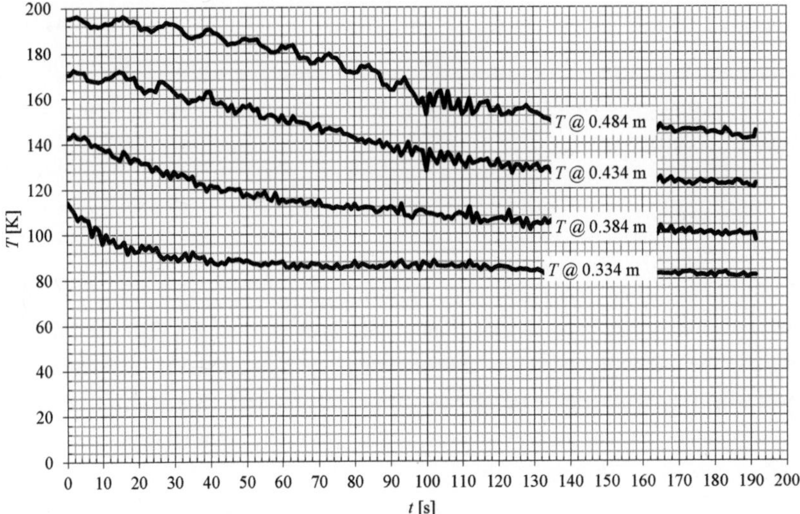

Figure 5-2. Numerical temperature development in the ullage vapour for a completely adiabatic system with no phase change and no heat exchange between liquid and vapour.

5.1.1 Limitations of the numerical method used in FLOW 3D

It will be shown here that all CFD programs based on the Volume of Fluid (VOF) method for the interface tracking and a single temperature per cell approximation will fail in simulating the experiments investigated here. The analysis that follows is based on information received via private communication with FLOW Science [34]. The analysis is also given in [35].

The VOF function defines the fraction of fluid (F) present in a cell. In case of a two fluid system it is 1.0 if fluid #1 completely fills the cell and 0 if fluid #1 is completely absent in the cell. The fraction of fluid is defined as:

$$F = \frac{V_1}{V}$$

where V_1 is the volume of fluid # 1 and V is the volume of the cell. Fluid properties in a cell are calculated using this fluid fraction F. The internal energy per unit volume of a cell is calculated as follows:

$$U = F\rho_1 I_1 + (1-F)\rho_2 I_2 \qquad\qquad 5\text{-}1$$

where $I = cT$. The average temperature T in a computational cell is calculated such that the internal energy is conserved:

$$T = \frac{\rho_1 c_1 T_1 V_1 + \rho_2 c_2 T_2 V_2}{\rho_1 c_1 V_1 + \rho_2 c_2 V_2} \qquad\qquad 5\text{-}2$$

From this equation it can be seen that if one fluid is much heavier than the other, the temperature in the cell will tend to the temperature of the heavier fluid. This degrades the accuracy at which the temperature of the other fluid is represented. It can also lead to large errors in the temperature when a flow is present. This is exactly what happens at the liquid-vapour interface in the CFD model.

To illustrate this, a simplified slosh model is considered (see Figure 5-3).

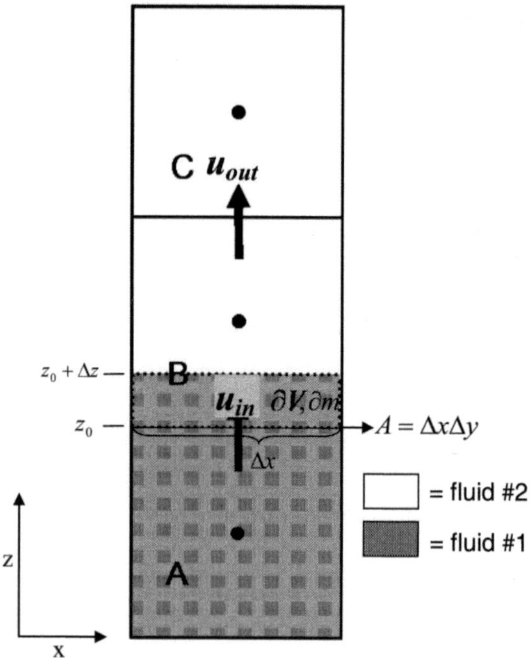

Figure 5-3. Simplified slosh model.

Assuming the fluid is incompressible equation 2-17 yields:

$$u\big|_{z0} = u\big|_{z0+\Delta z} = u_{in} \qquad\qquad 5\text{-}3$$

The mass of fluid #1 which enters cell B is expressed by $\partial m_{in} = \rho_1 u_{in}\Delta x\Delta y\partial t = \rho_1 u_{in}A\partial t$, where A is the surface area between two adjacent cells. The volume entering cell B is then found by:

$$\partial V_{in} = \frac{\partial m_{in}}{\rho} = u_{in}A\partial t \qquad\qquad 5\text{-}4$$

The *fractional* volume which flows into and out of fluid cell B can be expressed by:

$$\partial V_{fr\,in} = u_{in}\frac{A}{V}\partial t$$
$$5\text{-}5$$
$$\partial V_{fr\,out} = u_{out}\frac{A}{V}\partial t$$

If fluid #2 is also assumed incompressible the volume flowing into and out of cell B will be equal;

$$\partial V_{fr} = \partial V_{fr\,in} = \partial V_{fr\,out} \qquad\qquad 5\text{-}6$$

As shown in equation 5-2, the temperature in a cell is averaged such that internal energy is conserved. To shows that this can lead to errors, the following analysis will concentrate on the internal energy only, neglecting heat conduction and kinetic energy. The internal energy transported into and out of a cell is:

$$\partial U_{in} = \partial V_{fr}U_A$$
$$5\text{-}7$$
$$\partial U_{out} = \partial V_{fr}U_B$$

where $U_A = \left(F_A\rho_1 c_1 + (1 - F_A)\rho_2 c_2\right)T_A$ and $U_B = \left(F_B\rho_1 c_1 + (1 - F_B)\rho_2 c_2\right)T_B$

In the initial case that $F_A=1$ and $F_B=0$, the fluid fraction in cell B changes according to:

$$F_B^{new} = F_B^{old} + \partial V_{fr} F_A - \partial V_{fr} F_B = 0 + \partial V_{fr} - 0 = \partial V_{fr} \qquad \text{5-8}$$

The internal energy in cell B changes according to (using equations 5-1 and 5-7):

$$U_B^{new} = U_B^{old} + \partial V_{fr} U_A - \partial V_{fr} U_B = \left(F_B^{new} \rho_1 c_1 + (1 - F_B^{new}) \rho_2 c_2 \right) T_B^{new} \qquad \text{5-9}$$

By applying equation 5-9 the new temperature in cell B can be calculated:

$$T_B^{new} = \frac{U_B^{new}}{F_B^{new} \rho_1 c_1 + \left(1 - F_B^{new}\right) \rho_2 c_2} \qquad \text{5-10}$$

The equations above can be used to make an error estimation of the single temperature method.

For a worst-case error analysis it has to be assumed that:

$$F_A = 1$$
$$F_B = F_C = 0$$
$$T_B = T_C = T_2 \neq T_A = T_1$$

Using equations 5-9 and 5-10 and substituting the new values for the worst-case error analysis, the new temperature in cell B can be calculated:

$$T_B^{new} = \frac{\rho_2 c_2 T_2 + \partial V_{fr} \left(\rho_1 c_1 T_1 - \rho_2 c_2 T_2 \right)}{\partial V_{fr} \rho_1 c_1 + \left(1 - \partial V_{fr}\right) \rho_2 c_2} \qquad \text{5-11}$$

If fluid 1 is a liquid and fluid 2 is a gas then for most cases $\dfrac{\rho_2 c_2}{\rho_1 c_1} \ll 1$. For nitrogen this ratio is about 0.003 at 1 bar and $T_2 = 90$ K. The ratio gets even smaller for higher T_2. Because the ratio is so small it is possible to take the limit of equation 5-11 of $\rho_1 c_1 \to \infty$. By assuming that ∂V_{fr} is not too small this results in:

$$\underset{\rho_1 c_1 \to \infty}{LIM} \; \frac{\rho_2 c_2 T_2 + \partial V_{fr} \left(\rho_1 c_1 T_1 - \rho_2 c_2 T_2 \right)}{\partial V_{fr} \rho_1 c_1 + \left(1 - \partial V_{fr}\right) \rho_2 c_2} = T_1 \approx T_B^{new} \qquad \text{5-12}$$

Situation 1

If the flow is then reversed (as is the case for sloshing) and fluid 1 disappears from cell B, then using equation 5-9 and 5-10 the new temperature in cell B can be calculated:

$$U_B^{new2} = U_B^{new} + \partial V_{fr} U_C - \partial V_{fr} U_B^{new} = \rho_2 c_2 T_B^{new2} \qquad \text{5-13}$$

With $U_B^{new} = T_B^{new}(\partial V_{fr}\rho_1 c_1 + (1 - \partial V_{fr})\rho_2 c_2)$, $U_C = \rho_2 c_2 T_2$ and $U_B^{new} = \partial V_{fr}\rho_1 c_1 T_B^{new}$, this results in:

$$T_B^{new2} = (1 - \partial V_{fr})T_B^{new} + \partial V_{fr}T_2 \approx (1 - \partial V_{fr})T_1 + \partial V_{fr}T_2 \qquad \text{5-14}$$

After the back and forward fluid motion, cell A still has its initial temperature T_1 and cell C still has its initial temperature T_2 but cell B has a temperature which is an average of the two. In the absence of heat transfer between the fluids, the temperature in cell B should be at its initial temperature.

If the slosh motion is repeated, the error in cell B will get even bigger and the error is spread to the neighbouring cells. The longer the slosh time and the more back and forward fluid motions take place, the bigger the error gets. Because pressure is a function of temperature this will result in unrealistic pressure changes.

Situation 2

If the flow is not reversed but instead again a volume ∂V_{fr} flows from cell A into cell B and assuming that $2\partial V_{fr} < 1$ in cell B so that only fluid#2 flows into cell C, the new temperature in cell B will remain at $T_1 \approx T_B^{new}$ but the temperature in cell C will now change. By using equation 5-11 and knowing that in this case only ρ_2 and c_2 flow into cell C (and as a result $\rho_1 = \rho_2$ and $c_1 = c_2$ can be inserted in the equation) the new temperature in cell C is:

$$T_C^{new} = T_2(1 - \partial V_{fr}) + T_1 \partial V_{fr} \qquad \text{5-15}$$

The temperature in cell C should have remained at T_2 because only fluid #2 has flown into cell C. Because pressure is a function of temperature this will result in unrealistic pressure changes.

5.1.2   Simplified FLOW 3D slosh model

To demonstrate the short comings of FLOW 3D occurring in the situation discussed above, a simplified FLOW 3D model will be used. This model consists of two fluids, with fluid #1 being liquid nitrogen and fluid #2 being gaseous nitrogen. Thermal conductivity in the fluids is set to 0 so no heat transfer takes place within one fluid or between the two fluids. The phase change model has been switched off and the entire system is adiabatic.

The numerical model is shown in Figure 5-4 in a x-z axis projection. The model is 1 cell thick in the y-direction. The left part of Figure 5-4 shows the fluid fractions and the right part fluid temperatures. In the left part, red represents fluid fraction equal to 1 (in this case fluid #1, liquid nitrogen) and blue a fluid fraction of 0 (fluid #2, gaseous nitrogen). In the right part it can be seen that the liquid is much colder than the gas, 77 K and 160 K respectively. Gravity points in the negative z-direction, causing the fluid to oscillate around the neutral axis at $z = 1.5$.

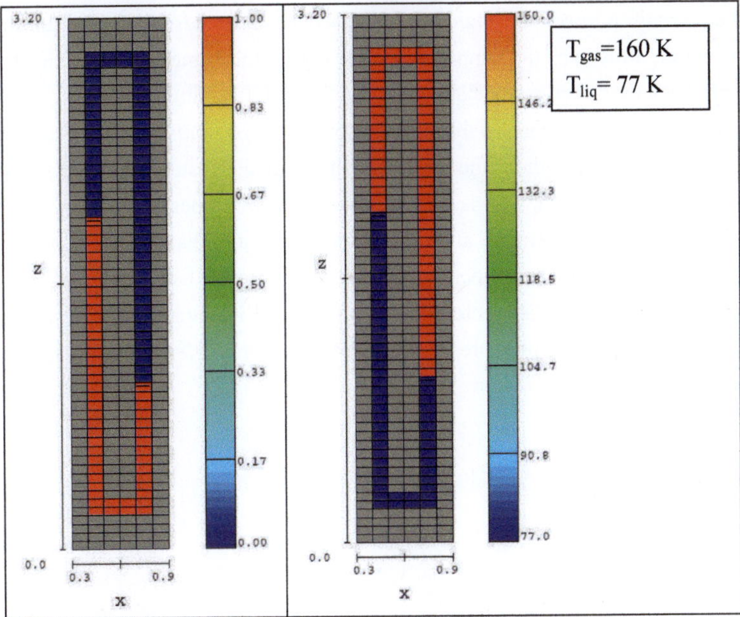

Figure 5-4. Numerical model used for the simplified analysis. The left part shows the fluid fraction and the right part shows the temperature (in K).

By zooming in at a cell which is completely filled with gas ($F = 0$) at $t = 0$ and which receives a flow from a cell filled with liquid ($F = 1$), an analysis of the temperature error can be made. This is illustrated in Figure 5-5. The top row shows the fluid fraction of certain cells at different times. The bottom row shows the temperature in the same cells at the same times. At $t = 0.0125$ s, a small volume of fluid #1 from cell A has flowed into cell B which previously only contained fluid #2. This causes the fluid fraction to rise from 0 to 0.1. The temperature in the cell immediately drops from 160 K to 78.9 K. At the same time, some volume of fluid #2 from cell B flows into cell C. Because the fluid which flowed into cell C has the same temperature as cell C, no temperature change in cell C is present. At $t = 0.01875$ s some more fluid from cell A flows into cell B, further increasing the fluid fraction in cell B. At the same time, some of fluid #2 from cell B flows into cell C. Because now the fluid #2 which flows from cell B has decreased in temperature, cell C also decreases in temperature. The temperature in cell C drops from 160 K to 159 K. This temperature drop of 1 K is too small to be visible in Figure 5-5. At $t = 0.0298$ the previous cycle is repeated, but now the temperature in cell C has dropped from 159 K to 155 K, which is clearly visible in the figure.

The temperature drop in cell C should not be present at all because it consists of only fluid #2. This example illustrates the temperature error due to the numerical approximation very well and has been described mathematically in section 5.1.1 by situation 2.

As time increases the error spreads through the complete system. Figure 5-6 shows how the temperature develops throughout the gaseous volume of the system. After 10 seconds the maximum temperature in the gas is 148 K. The average temperature in the gas is even lower. The error in the temperature after 10 s is therefore more than 12 K. Because pressure in the system is determined as a function of the temperature (by using the ideal gas model), the pressure in the system also decreases, as can be seen in Figure 5-7. The pressure has decreased from atmospheric pressure (0.103 MPa) to less than 0.8 MPa. This is an error of more than 20%.

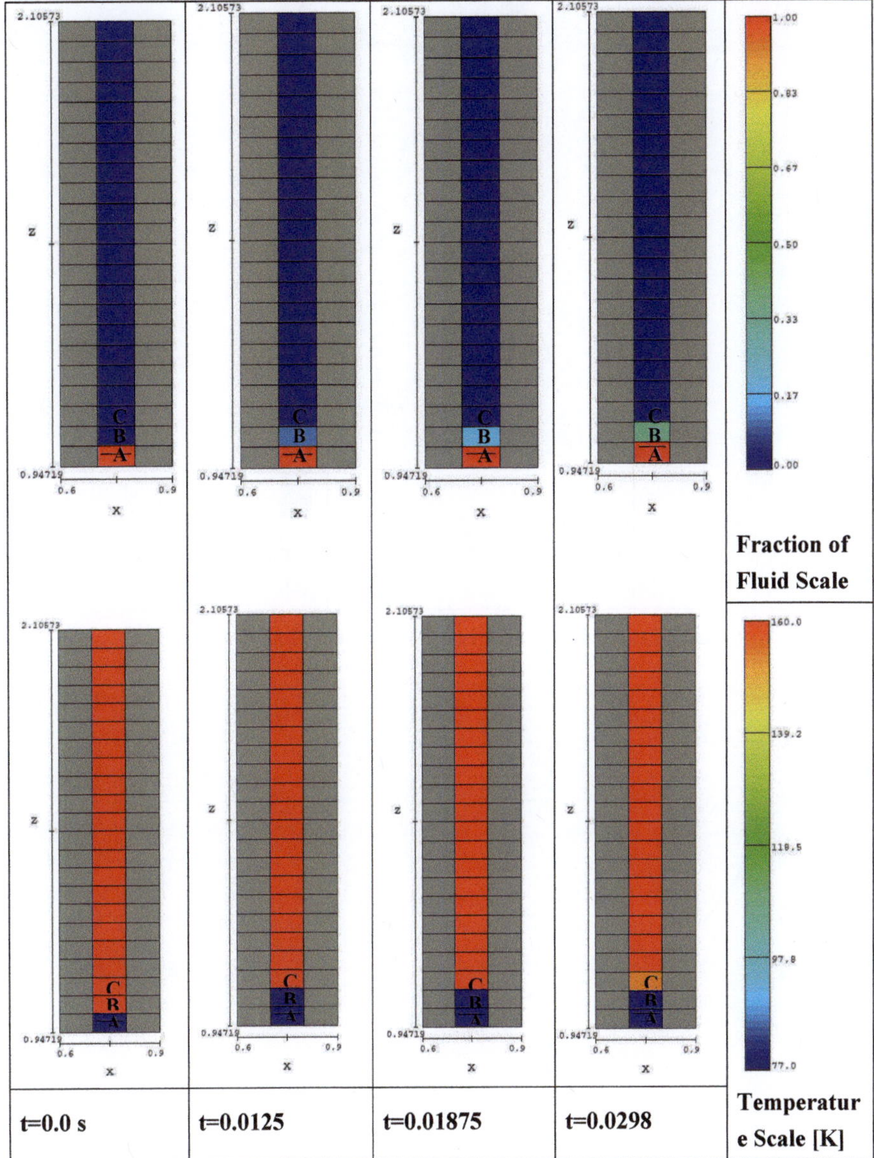

Figure 5-5. The temperature error in a cell caused by cold liquid which flows into a cell with hot gas. The top row shows the fraction of fluid and bottom row the resulting temperature. Time increases from left to right.

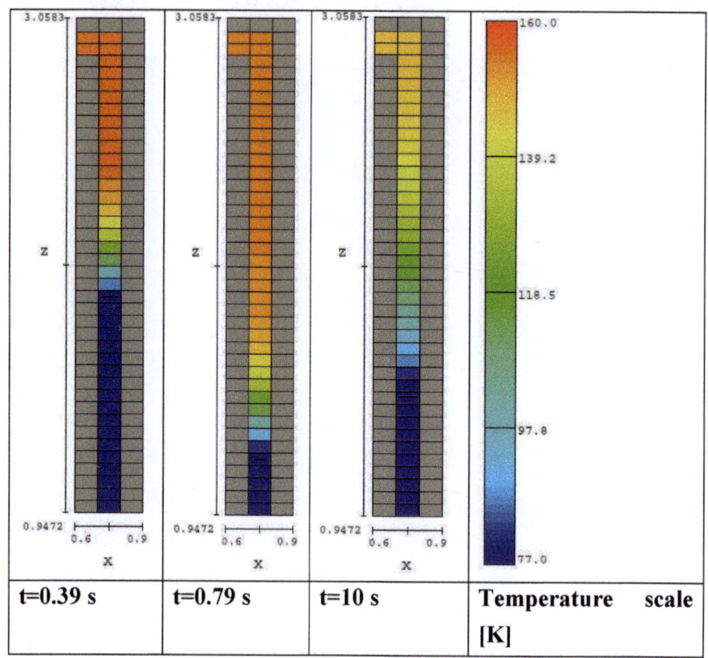

| t=0.39 s | t=0.79 s | t=10 s | Temperature scale [K] |

Figure 5-6. Increase of the temperature error with increasing time.

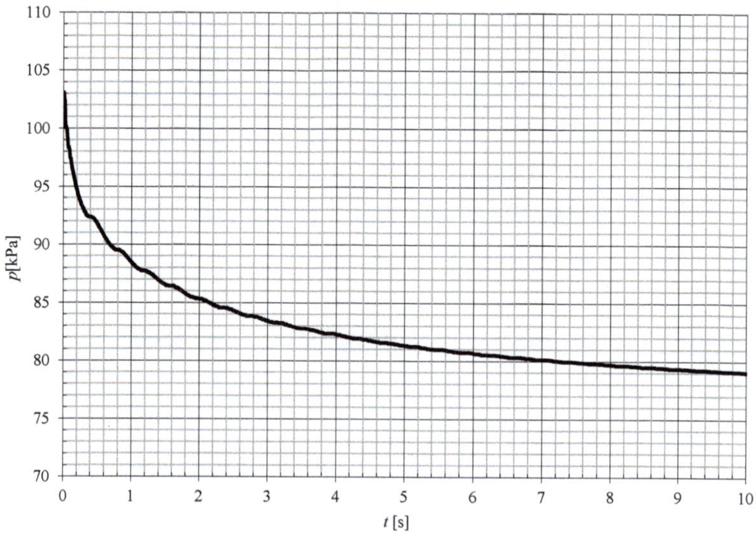

Figure 5-7. Pressure drop in the simplified FLOW 3D model due to the numerical error in the temperature.

5.1.3   Possible solutions for the error

First of all refining the grid around the free liquid surface will not solve the problem. The finer mesh increases the computational time and will not get rid of the root cause of the error. A finer mesh would slow down the spreading of the error through the system, but for longer durations it would still grow to unacceptable proportions.

In private communication discussing the error, FLOW Science has proposed two possible solutions to the problem [34]. One is to move completely away from using a single temperature per cell. Two temperatures would be used in each cell, one for the temperature of fluid #1 and the other the temperature of fluid #2. Accordingly, two internal energies would have to be calculated in each cell to represent each fluid. An additional term accounting for the heat transfer between the two fluids would also have to be added.

A second, simpler approach could also be taken. In this case the single temperature approach would be maintained, but the advective calculation would be modified. By assuming that in any cell containing both fluid #1 and #2, the mixture temperature is dominated by the liquid in the cell it therefore should not be used for computing internal energy flowing into the cells that contain only gas. The internal energy of gas which flows from a cell containing liquid into a pure gas cell would have to be modified to avoid using the liquid-dominated temperature in the donor cell, for example, by using the acceptor cell temperature.

As soon as the flow is reversed and when the liquid has completely flown out of a cell and replaced by gas, the cell temperature would have to be reset to a representative gas temperature (for example using the temperature of its fully gas neighbours) because the old 'liquid' temperature is not representative of the newly entered gas.

5.2 A 1D model for the simulation of heat and mass transfer during sloshing

After failure of the FLOW 3D model to accurately simulate the pressure drop due to sloshing the attention was turned to the 1D stratification model developed in chapter 4, section 4.4.

To this end a 1D energy balance will be set up for a sloshing liquid. It will be shown that only minor adaptations to the 1D stratification model developed in chapter 4 are needed. These are:

1. A Nusselt number is used to adapt the liquid conduction coefficient by replacing λ_l with $\lambda_{l,eff}$ in the liquid region affected by the sloshing motion (discussed in section 5.2.1-5.2.4)

2. The value for $\dot{Q}_{l-n}$ has to be increased in the liquid region affected by the sloshing motion (discussed in section 5.2.5)

5.2.1 Energy balance

As discussed in chapter 4 a thermal gradient in the liquid is created by heat input into the liquid at the liquid surface. As the temperature in the top liquid layer increases, a conductive heat flow through the liquid will cause the lower liquid regions to heat up as well and the thickness of the stratified layer will increase over time. A one dimensional energy equation can be set up for the thermal gradient in the liquid, based on equation 2-21:

$$\lambda_l \frac{\partial^2 T_l}{\partial z^2} = \rho_l c_{p,l}\left(\frac{\partial T_l}{\partial t} + u_{z,l}\frac{\partial T}{\partial z}\right) \qquad 5\text{-}16$$

which is equation 4-1 but including the term describing convective heat transfer. The boundary condition at the liquid surface is described by equation 2-22:

$$\dot{Q}_{pch} = S_{ls}\tilde{L}\rho_l(u_{z,l} - u_{z,\Gamma}) = \tilde{L}\,\dot{m}_{pch} = S_{ls}\lambda_v \left.\frac{\partial T_v}{\partial z}\right|_{ls} - S_{ls}\lambda_l \left.\frac{\partial T_l}{\partial z}\right|_{ls} \qquad 5\text{-}17$$

where $\left.\dfrac{\partial T_l}{\partial z}\right|_{ls}$ is approximated by $\left.\dfrac{\partial T_l}{\partial z}\right|_{ls} = \dfrac{\Delta T_l}{\delta_T}$ and $\left.\dfrac{\partial T_v}{\partial z}\right|_{ls}$ has been determined in section 4.1.

5.2.2 The energy balance for sloshing liquids

Sloshing will cause convective currents in the liquid which will increase the heat flow through the liquid. The convection term in equation 5-16 is not zero anymore. According to [29] the convective term in turbulent boundary layers can be modelled as an equivalent conductive term, $u_z \dfrac{\partial T}{\partial z} = \dfrac{-\lambda_{conv}}{\rho c_p} \dfrac{\partial^2 T}{\partial z^2}$. In [20] and [39] it is shown that this also applies for a sloshing liquid, so that equation 5-16 becomes:

$$\left(\lambda_l + \lambda_{l,conv}\right)\frac{\partial^2 T_l}{\partial z^2} = \lambda_{l,eff}\frac{\partial^2 T_l}{\partial z^2} = \rho c_{p,l}\left(\frac{\partial T_l}{\partial t}\right) \qquad \text{5-18}$$

where $\lambda_{l,eff}$ is an effective heat conduction coefficient.

The sloshing motion will also affect the heat transfer in the ullage, so λ_v is replaced by $\lambda_{v,eff}$ which leads to the following formulation of the boundary condition at the liquid surface (equation 2-22):

$$\dot{Q}_{pch} = S_{ls}\tilde{L}\rho_l(u_{z,l} - u_{z,\Gamma}) = \tilde{L}\,\dot{m}_{pch} = S_{ls}\lambda_{v,eff}\left.\frac{\partial T_v}{\partial z}\right|_{ls} - S_{ls}\lambda_{l,eff}\left.\frac{\partial T_l}{\partial z}\right|_{ls} \qquad \text{5-19}$$

It is convenient to write equation 5-19 as $S_{ls}\lambda_{l,eff}\left.\dfrac{\partial T_l}{\partial z}\right|_{ls} = S_{ls}\lambda_{v,eff}\left.\dfrac{\partial T_v}{\partial z}\right|_{ls} - \dot{Q}_{pch}$, where the 2nd and 3rd term represent heat extracted from the ullage. This can be determined by equation 4-5 but the equation now has to be solved for the heat flux. By replacing $\dot{Q}_v$ with $\dot{Q}_{conv}$ this yields:

$$\dot{Q}_{conv} = S_{ls}\,\lambda_{v,eff}\,\frac{\partial T_v}{\partial z}\bigg|_{ls} - \dot{Q}_{pch} = \dot{I}_v - \dot{m}_{pch}\,h_e =$$

5-20

$$= \dot{m}_{pch}\,c_v T_{v,av} + m_v\,c_v\,\dot{T}_{v,av} - \dot{m}_{pch}\,h_e .^{6\ 7}$$

The heat flow $\dot{Q}_{conv}$ is evaluated at maximum rate of pressure drop using the values presented below.

- The condensation mass flow rate $\dot{m}_{pch}$ is determined using $\dot{m}_{pch} = \dot{\rho}\,V_u$, where $\dot{\rho}$ can be determined using equation 3-2 and the ideal gas law $p = \rho\widetilde{R}T$ yielding $\dot{\rho} = \dfrac{\dot{p}}{RT} - \dfrac{p\dot{T}}{RT^2}$. Values used for the calculation are:

    - $V_u = 0.0228$ m^3 (V_u is determined from geometrical data in Figure 3-1);

    - $\dot{p} = -\left|\dot{p}\right|_{max}$ (from Table III-2);

    - $T = T_{v,av}\big|_{tpd}$ (from Table III-2);

    - $\dot{T} = \dot{T}_v\big|_{tpd}$ (from Table 3-2 for actively pressurized experiments, for self-pressurized experiments the value is zero as shown in section 3.2.1);

    - $\widetilde{R} = 300$ J kg^{-1} K^{-1}.

    The resulting values for $\dot{m}_{pch}$ are presented in Table III-2.

- The value for h_e can be looked up in NIST property tables (appendix VII) and is listed in Table III-2.

- The value for c_v can also be found in appendix VII and is 740 J kg^{-1} K^{-1}.

- The value for $T_{v,av}$ and $\dot{T}_{v,av}$ are taken identical to T and $\dot{T}$, which were discussed above.

---

[6] This equation can also be written as $\dot{m}_{pch}(c_v T_{ls} - h_e) + \dot{m}_{pch}\,c_v(T_{v,av} - T_{ls}) + m_v\,c_v\,\dot{T}_{v,av}$, where the first term corresponds to the latent heat.

[7] The dewar is excited during sloshing so this involves a certain work. However this work is converted directly into kinetic energy so the terms cancel each other out, resulting in the equation presented here.

Using the values presented above $\dot{Q}_{conv}$ can now be determined. The result is listed in Table III-2.

5.2.3   The Nusselt number in sloshing liquids

The ratio between $\lambda_{l,eff}$ and λ_l represents a Nusselt number such that $\lambda_{l,eff} = \text{Nu}\lambda_l$. This Nusselt number is of particular interest as it is an important input for the 1D slosh model. This section will describe how the value of Nu can be determined.

By substituting equation 5-20 in equation 5-19 and rearranging terms the following relation is obtained:

$$\lambda_{l,eff} S_{ls} \left.\frac{\partial T_l}{\partial z}\right|_{ls} = \dot{Q}_{conv} \qquad\qquad\qquad 5\text{-}21$$

By using the relation $\lambda_{l,eff} = \text{Nu}\lambda_l$ and by expressing the thermal gradient in the liquid at t_{pd}

as $\left.\dfrac{\partial T_l}{\partial z}\right|_{ls} = \left.\dfrac{\Delta T_l}{\delta_T}\right|_{tpd}$ equation 5-21 can be written as:

$$\text{Nu}\lambda_l S_{ls} \left.\frac{\Delta T_l}{\delta_T}\right|_{tpd} = \dot{Q}_{conv} \qquad\qquad\qquad 5\text{-}22$$

The value of ΔT_l can be obtained from the experimental data by subtracting the temperature in the bulk of the liquid from the temperature at the liquid surface. In Figure 5-8 this is shown for sp120 and sp160. Table III-2 gives the value of ΔT_l for all experiments.

In case of the self-pressurized experiments the thickness δ_T of the thermal layer in the liquid can also be determined from the experimental temperature data, see Figure 5-8 for sp120 and sp160. But for the actively pressurised experiments thickness of the thermal layer is so small that it is not captured accurately by the temperature sensors, so another method must be applied to find δ_T.

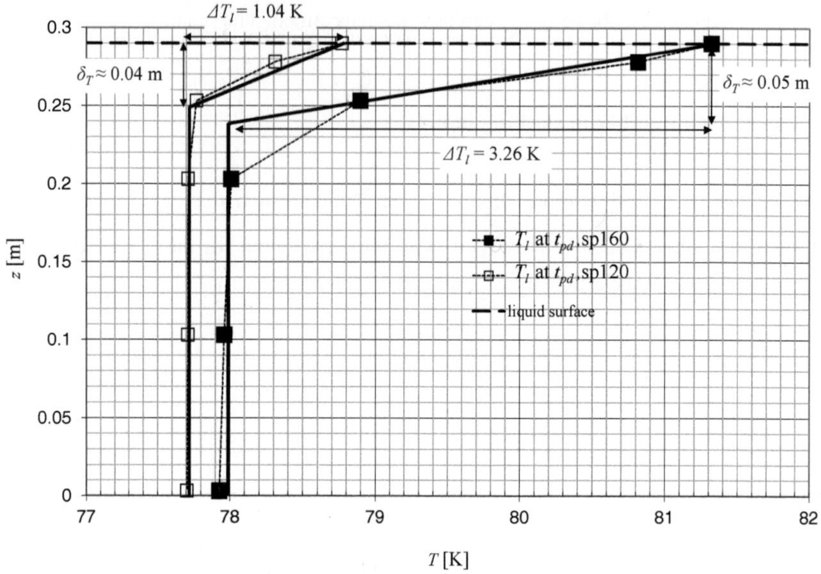

Figure 5-8. Temperature in the liquid at t_{pd} for the experiments sp160 and sp120.

Alternatively δ_T can be determined with equation 2-71. Because δ_T is evaluated at t_{pd} equation 2-71 has to be rewritten in:

$$\delta_T\big|_{tpd} = C\left(\sqrt{\alpha t_{press} + \mathrm{Nu}\alpha t_{tpd}}\right)$$

5-23

The parameter α in this equation represents the thermal diffusivity according to equation 2-33. Its value is determined by using $\lambda_l = 0.15$ W m^{-2}, $\rho_l = 805$ kg m^{-3} and $c_p = 2040$ J kg^{-1} K^{-1} from appendix VII, yielding $\alpha = 9.1$E-08 m^2 s^{-1}. The values for t_{press} and t_{pd} can be found in Table III-1 and Table III-2 respectively.

In [20] it is proposed to use $C = 3$, but here a value of $C = 2.5$ will be used. This value was found by calculating δ_T at t_s for sp160, using $\delta_T\big|_{ts} = C\left(\sqrt{\alpha t_{press}}\right)$. This yields $\delta_T\big|_{ts} = 0.046$ m with $C = 2.5$, $\alpha = 9.1$E-08 m^2 s^{-1} and $t_{press} = 3720$ s. For $C = 3$, $\delta_T\big|_{ts} = 0.055$m is found. Then the found values were compared to the results with the measured gradient at t_s presented in Figure 5-9. A value of $C = 2.5$ reflects the experimentally measured thickness better than $C = 3$.

By substitution of equation 5-23 into 5-22 the Nusselt number can be determined:

$$\mathrm{Nu}\lambda_l S_{ls} \frac{\Delta T_l\big|_{tpd}}{2.5\sqrt{\alpha t_{press}} + \mathrm{Nu}\alpha t_{tpd}} = \dot{Q}_{conv}$$

$$5\text{-}24$$

Equation 5-24 can now be solved for Nu by using $S_{ls} = \pi R^2$ with $R = 0.145$ m (the radius of the liquid surface). Applying this quadratic equation to the experiments yields a negative and a positive solution, of which only the positive solution makes physical sense.

Then, equation 5-23 can be applied to find $\delta_T\big|_{ts}$. Results are presented in Table III-2. The Nusselt numbers are also depicted in Figure 5-10. The Nusselt numbers are approximately the same for each experiment, with an average value of $\mathrm{Nu_{av}} = 44$. The shaded band indicates the error of $\mathrm{Nu_{av}}$.

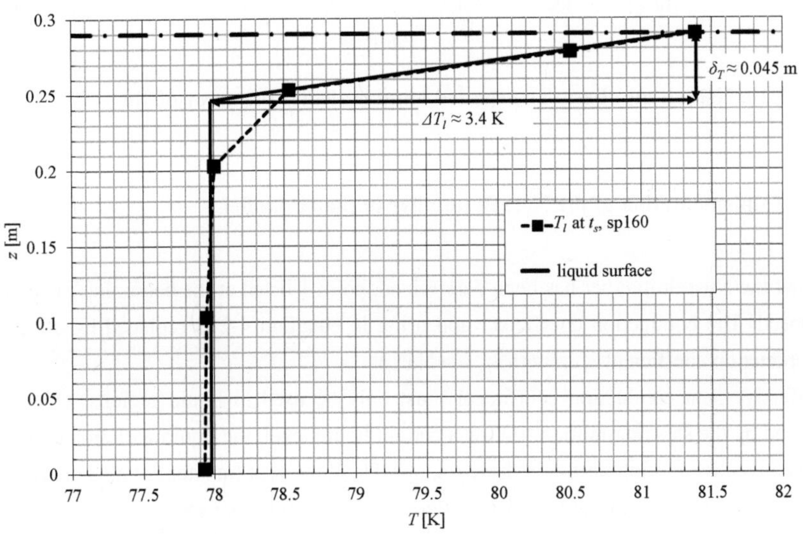

Figure 5-9. Temperature in the liquid at t_s for the experiments sp160.

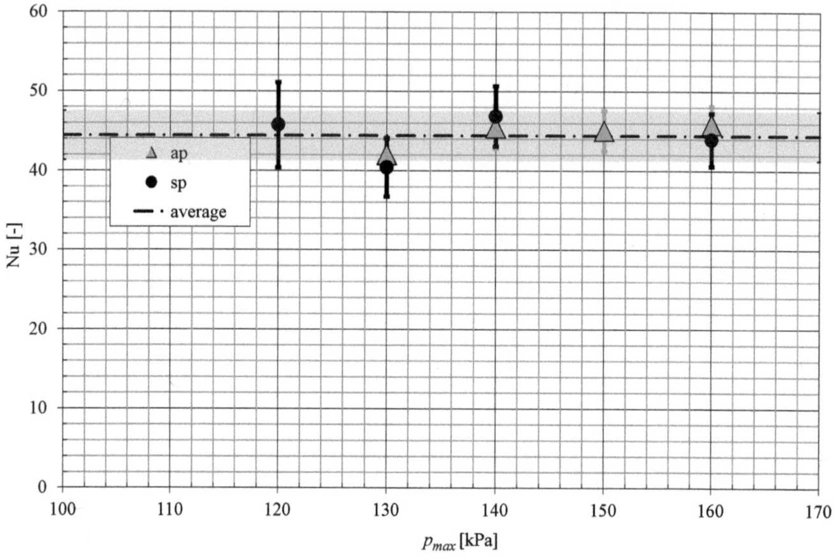

Figure 5-10. Nusselt numbers. The shaded band indicates the error of Nu_{av} .

5.2.4   Implementation of the Nusselt number in the 1D model

Now that the Nusselt number has been determined, point 1 listed at the beginning of section 0 is closed. The conduction coefficient appearing in equation 4-2 is replaced by effective heat conduction coefficient $\lambda_{l,eff}$ (with $\lambda_{l,eff} = Nu\,\lambda_l$).

The effective heat conduction coefficient must only be used in liquid regions which are influenced by the sloshing motion. Below a certain liquid level, the liquid will remain at rest and no convective motion occurs in these regions. Here, the heat transfer remains unaffected by the sloshing and the normal heat conduction coefficient must be used.

Therefore it is of importance that before running a simulation the region of influence h_{infl} is determined.

In the experiments discussed here the region of influence is assumed to be the liquid region above the dome; $z > 0.145$ m. This assumption is based on the fact that for fill levels of $\dfrac{H_f}{R} > 1$ the dome has no influence on damping anymore (see Figure 2-8). This suggests that the liquid in the dome remains at rest during sloshing. This is supported by the fact that the two temperature sensors in the dome, T14 and T15, indicate a homogeneous temperature distribution unaffected by the sloshing motion (see for example Figure 3-14 and Figure 3-15).

The region of influence is then defined as the height of the liquid region influenced by the sloshing, h_{infl} = 0.29 m - 0.145 m = 0.145 m. Thus $\lambda_{l,eff}$ is used if $z > h_{fill}\text{-}h_{infl}$, and λ_l if $z < h_{fill}\text{-}h_{infl}$ is inserted in the 1D model.

5.2.5 Heat input during sloshing

Point 2 listed at beginning of section 0 will be elaborated on in this section.

From Table 3-3 it is known that about 6.7 W of heat enters the liquid ($\dot{Q}_l$) during the pressurization phase. In section 5.2.3 it has been shown that during sloshing an increase in heat is transfer is observed from ullage to liquid. It is therefore expected that over a given time interval Δt the heat which enters the liquid can be written as:

$$Q_{l,slosh} = \dot{Q}_l\,\Delta t + \dot{Q}_{conv,av}\,\Delta t \qquad\qquad 5\text{-}25$$

where $\dot{Q}_{conv,av}$ is the average heat flow from ullage to liquid over the specified time interval, determined using equation 5-20 in differential form as described below.

The time between slosh initiation (t_s) and minimum pressure during sloshing (t_{pmin}) is chosen as the time interval, $\Delta t = t_{pmin} - t_s$. Equation 5-20 can then be written as

$$Q_{conv} = \dot{Q}_{conv,av}\,\Delta t = I_v\Big|_{ts}^{tpmin} - m_v\Big|_{ts}^{tpmin} h_e = m_v c_v T_{v,av}\Big|_{ts}^{tpmin} - m_v\Big|_{ts}^{tpmin} h_e .$$

The vapor mass m_v at t_s and the temperature $T_{v,av}$ at t_s are taken from Table III-1. The temperature at minimum pressure $T_{v,av}$ is taken from Figure 3-6 and Figure 3-11 using the time of minimum pressure defined in Figure 3-4 and Figure 3-9. These temperatures are listed in Table 5-1. The vapor mass at minimum pressure is determined using these temperatures and the ideal gas equation and the ullage volume V_u (specified in section 5.2.2); $m_v\big|_{tpmin} = \dfrac{p_{tpmin}}{\tilde{R}\,T_{v,av}\big|_{tpmin}} V_u$, with

$\tilde{R}$ = 300 J kg^{-1} K^{-1} and p_{min} as defined in Figure 3-4 and Figure 3-9. Results for the vapour mass are again listed in Table 5-1. The values of c_v and h_e are specified in section 5.2.2.

Finally, this results in $\dot{Q}_{conv,av}$ specified in Table 5-1.

Using equation 5-25, $Q_{l,slosh}$ can be determined.

This value can be crosschecked by calculating the internal energies of the liquid at slosh initiation and at minimum pressure during slosh (using the method in section 3.4.1). Using this method a much larger value for the heat flow is found, so that equation 5-25 has to be rewritten as:

$$Q_{l,slosh} = \dot{Q}_l\,\Delta t + \dot{Q}_{conv,av}\,\Delta t + \dot{Q}_{add}\,\Delta t \qquad\qquad 5\text{-}26$$

where $\dot{Q}_{add}$ is the additional heat flow into the liquid. Values can be found in Table 5-1.

An explanation for the increased heat flux might be that the sloshing liquid comes into contact with the warmer ullage wall. Additionally the thermal boundary layer at the dewar wall may be destroyed by the sloshing, resulting in an increase of the heat which enters the liquid through the wall.

This increased heat flow must be taken into account in the 1D model. In the 1D model, the heat $\dot{Q}_{add}$ is only added to the liquid in the region of influence because only in this region the thermal boundary layer will be affected.

Coming back to the 1D model as described in section 4.4.1 this means that the correct value for $\dot{Q}_{l-n}^{2,k}$ is thus obtained by increasing $\dot{Q}_{l-n}^{2,k}$ with $\dot{Q}_{add}/n_{infl}$ in those nodes that lie in the region of influence.

It is noted that for applications to full scale tanks, $\dot{Q}_l$ and $\dot{Q}_{add}$ will probably be less important factor in the 1D slosh model because of the much lower tank area to tank volume ratio.

	Δt	$T_{v,av}\big|_{tpmin}$	$m_v\big|_{tpmin}$	$\dot{Q}_{l,slosh}$	$\dot{Q}_{conv,av}$	$\dot{Q}_l$	$\dot{Q}_{add}$
		[K]	[kg]	[W]	[W]	[W]	[W]
sp160	688	165	0.0611	18.5	3.3	6.7	8.5
sp120	220	165	0.0527	17.4	1.8	6.7	8.9
ap160	487	162	0.0554	18.8	6.8	6.7	5.3
ap130	218	162.5	0.0526	17.8	4.3	6.7	6.8

Table 5-1. Heat input into the liquid during sloshing.

5.3   Results of the 1D slosh model

After adapting the 1D model presented in section 4.4.1 according to the discussion in section 0 the model was applied to the experiments. The results are discussed below. As an initial condition for the slosh model the liquid temperature in each layer is preset according to the thermal stratification present in the experiments at slosh initiation (according to Figure 3-14 and Figure 3-15). Other input was as specified in section 4.4.2.

Figure 5-11 shows the pressure development as predicted by the 1D model compared to experimental results for experiments sp160 and sp120. Figure 5-12 shows this for ap160 and ap130. Minimum pressure and time of minimum pressure predicted by the model compared with the experiments are listed in Table 5-2. From the table and figures it can be seen that the 1D model predicts a somewhat lower minimum pressure. The time of minimum pressure occurs later than during the experiments.

		t_{pmin} [s]	p_{min} [kPa]
sp160	experiment	688	134.5
	1D model	784	132.3
sp120	experiment	220	114.2
	1D model	259	113.6
ap160	experiment	487	121.9
	1D model	744	120.6
ap130	experiment	218	114.6
	1D model	381	113.9

Table 5-2. Comparison of p_{min} and t_{pmin} between 1D model and experimental values.

After minimum pressure has been reached, the pressure starts to rise slowly again. This pressure rise is also captured by the 1D model. The thermal gradient in the liquid has become small and the heat transported through the liquid is smaller than the heat entering the liquid from the surroundings. Consequentially the liquid at the surface starts to heat up and evaporation occurs which leads to a pressure increase.

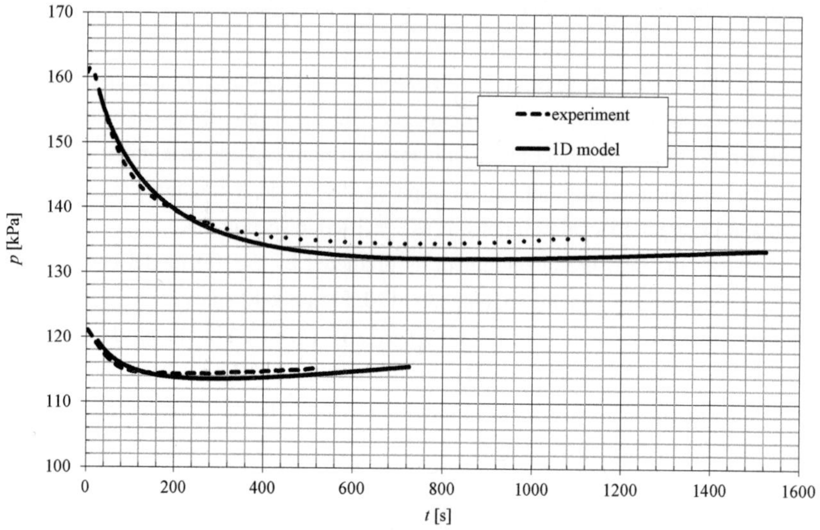

Figure 5-11. Pressure development according to the 1D model compared to experiment sp160 and sp120.

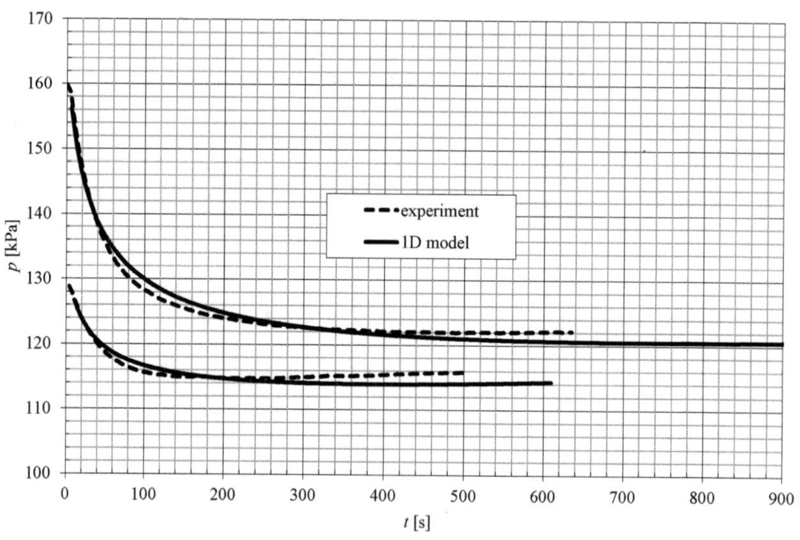

Figure 5-12. Pressure development according to the 1D model compared to experiment ap160 and ap130.

Temperature distribution in the liquid at minimum pressure ($t=t_{pmin}$) predicted by the 1D model is compared to the results for the self-pressurisation experiments in Figure 5-13 and for the actively pressurized experiments in Figure 5-14. The temperature according to the 1D model is in good agreement with the experimental results.

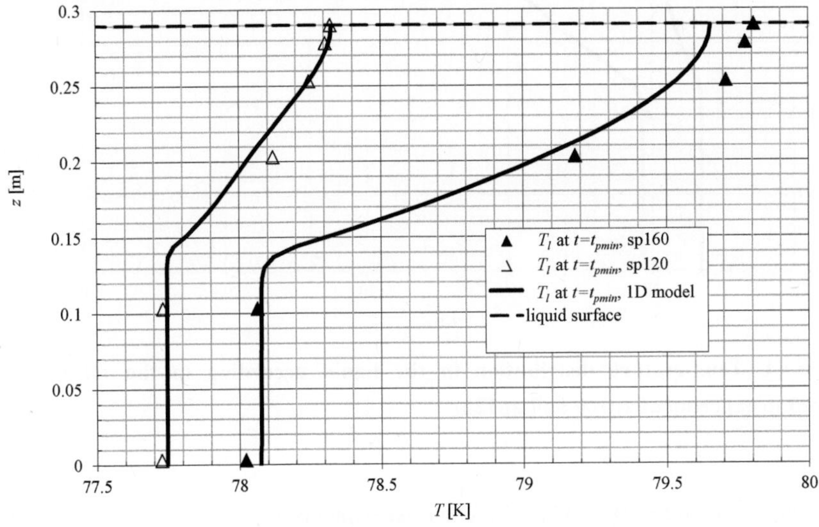

Figure 5-13. Temperature distribution in the liquid according to the 1D model compared to experiment sp160 and sp120.

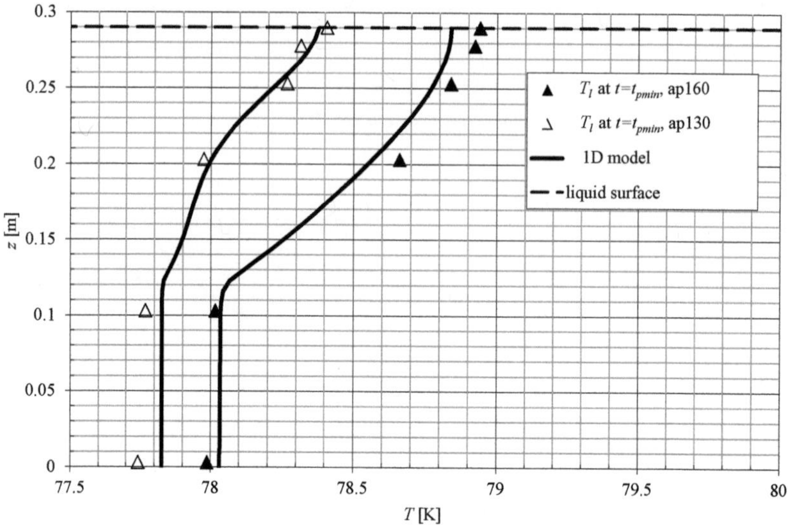

Figure 5-14. Temperature distribution in the liquid according to the 1D model compared to experiment ap160 and ap120.

5.4 Comparison of the 1D model with other experiments

Data from experiments conducted by Lacapere et al. [19] and Moran et al. [18] are used to make further comparisons with the predictions made by the 1D model. The experiments are also described briefly in section 1.1.

One of the experiments by Lacapere at al. was done using oxygen. The dewar was pressurised from atmospheric pressure to 250 kPa in 100 s.

Moran et al. used Hydrogen. Moran et al. have published data on 19 experiments in which the tank was pressurized with hydrogen gas. For each of the experiments the maximum rate of pressure drop has been determined, but for only very few experiments the complete pressure response over time is published. One of the experiments where the complete pressure response is known is the experiment named RDG 869. During this experiment, the tank was pressurized to 245 kPa in 41 s.

Table 5-3 gives fluid properties for both oxygen and hydrogen.

	p_{sat} [kPa]	T_{sat} [K]	λ_l [W m-1K-1]	$c_{p,l}$ [J kg-1K-1]	$c_{v,l}$ [J kg^{-1}K^{-1}]	ρ_l [kg m-3]	α [m2 s-1]	μ [μPa•s]	h_e [J kg-1]
O_2	250	99.81	0.138	1737	896	1092	7.3E-8	153	-116780
H_2	250	23.86	0.105	12194	5881	66.4	1.3E-7	10.2	39224

Table 5-3. Fluid data for oxygen and hydrogen at $p = 250$ kPa and saturation conditions.

Based on their data, information for setting up the 1D slosh model can be extracted. First of all $\dot{Q}_{conv}$ is determined according to equation 5-20. The necessary input extracted from the publications is given in Table 5-4. The average ullage temperature has been estimated based on experimental data. Note that for Hydrogen vapour the approximation $i = c_v T$ is not valid so the internal energy must directly taken from property tables. Its value is given in Table 5-4 for $p = 2.5$ bar and $T = 31.2$ K. Under these conditions $\tilde{R} = 5335$ J kg^{-1} K^{-1}. For LOX the internal energy is also provided directly in the table for $p = 2.5$ bar and $T = 138$ K. Under these conditions $\tilde{R} = 284$ J kg^{-1} K^{-1}. No ullage temperature evolution during sloshing is available so $\dot{T}_{v,av}$ is set to zero. The ullage volume V_u and liquid surface area S_{ls} have been determined from geometric data specified in the publications (see also section 1.1). The results are listed in Table 5-4.

The Nusselt number is determined by using equation 5-24. ΔT_i is determined by subtracting the saturation temperatures at the beginning and end of the pressurisation phase, and is listed in Table 5-4. It is assumed that maximum rate of pressure drop occurs immediately after slosh initiation so $t_{pd} = 0$ s. The thermal diffusion coefficient α is listed in Table 5-3.

In Table 5-4 it can be seen that for the experiment of Lacapere et al. the Nusselt number is much higher than for the all the other experiments which have been analysed so far. This is explained by the fact that Lacapere et al. deliberately introduced a slosh motion at the eigenfrequency of the system, resulting in very violent sloshing.

	$T_{v,av}$	i	t_{press}	$-\left\|\dfrac{dp}{dt}\right\|_{max}$	ΔT_l	tpd	δT	Vu	Sls	$\dot{Q}_{conv}$	Nu
	[K]	[J kg^{-1}]	[s]	[kPa/s]	[K]	[s]	[m]	[m^3]	[m^2]	[W]	[-]
Lacapere (LOX)	138	87769	100	-14.0	9.6	0	7E-3	8.3E-3	0.026	606	119
Moran (RDG 869)	31.2	435050	41	-0.76	3.26	0	6.2E-3	0.58	1.66	1043	11

Table 5-4. Data for determination of the Nusselt number of other experiments.

Using the information given above the 1D slosh model discussed earlier in this chapter can be applied. There is no information on the heat entering the liquid through the dewar walls so $\dot{Q}_{l-n}+\dot{Q}_{add}$ is set to 0. It is expected that the experiments of Moran et al. will be less sensitive to errors introduced by this assumption because the dewar dimensions are relatively large. This results in a lower area to volume ratio.

For the experiments of Moran et al. the region of influence is determined by calculating the pendulum length L_l for spherical tanks from [6]. This yields $h_{infl} = L_l = 0.41$ m. Because the mechanical model is not valid for chaotic sloshing near the eigenfrequency of the system, it cannot be used in case of the experiments of Lacapere et al. In this case it is estimated that $h_{infl} = 0.15$ m, based on their numerical results in [19] where it can be seen that below a certain level the liquid temperature remains unaffected by the sloshing motion.

Results are displayed in Figure 5-15 and Figure 5-16. Figure 5-15 shows the development of pressure during sloshing for the experiment by Lacapere et al. and Figure 5-16 shows the results for experiment RDG 869 by Moran et al. It must be noted that the experimental data was extracted from graphs presented in the article, and this leads to an error estimated at $\pm$ 2 kPa.

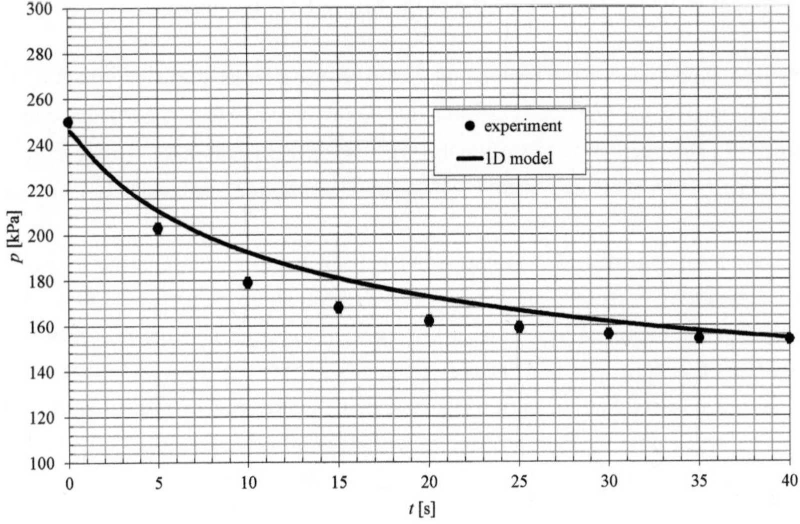

Figure 5-15. Pressure development for the test of Lacapere et al using LOX.

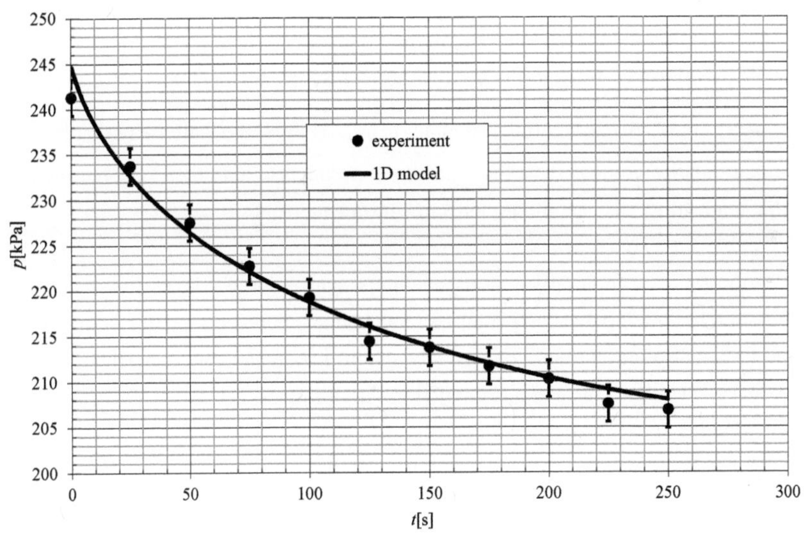

Figure 5-16. Pressure development for experiment RDG 869 of Moran et al.

5.5   Application of the 1D slosh model to the future ESC-B upper stage

Now that the experiments have been simulated in an acceptable manner by the 1D slosh model it is of interest to apply the 1D slosh model to the future ESC-B upper stage, depicted in Figure 5-17. This section will focus on the pressure development in the hydrogen tank of the stage. The results presented in this section are also provided in tabular form in Table III-12.

As there is no experimental data available the Nusselt number cannot be determined, so instead analyses have been done for three different Nusselt numbers, namely 1, 40 and 80. The volume of the hydrogen tank is 64 m^3. According to Arndt [42] the hydrogen ullage factor at lift off is about 5%, which yields $V_u = 3.2$ m^3.

Furthermore, it is stated that the pressurisation time of the hydrogen tank before lift-off is 3600 s, after which time the tank pressure is $p = 330$ kPa. Using equation 5-23 with α as specified in Table 5-3 and $t_{pd} = 0$ yields a thermally stratified thickness of $\delta_T = 0.054$ m. At 100 kPa, $T_{sat} = 20.2$ K. At 330 kPa, $T_{sat} = 25$ K so that $\Delta T_l = 4.8$ K.

This thermal gradient is set as an initial condition of the 1D model, also indicated by the dotted line in Figure 5-19.

There is no information on the ullage temperature, so this value is set to the value used in the simulation of experiment RDG 869 by Moran et al. This yields $T_v = 31.2$ K.

There is also no information regarding the heat input through the tank wall, so it is assumed that $\dot{Q}_{l-n} + \dot{Q}_{add} = 0$. Because a full scale tank is treated here, the wall area to tank volume ratio is much smaller compared to those in the experiments. It is therefore reasonable to assume that neglecting this heat input yields insignificant errors.

Because of the rather complex hydrogen tank shape with a large convex lower dome it is assumed that the liquid between the top of the convex dome up to the liquid level is influenced by the sloshing motion. This is indicated in Figure 5-17 by h_{infl}, where $h_{infl} \approx 1.9$ m. The radius of the liquid surface is estimated at $R_{ls} = 1.8$ m.

In Figure 5-18 the resulting pressure change over 120 s is shown for different Nusselt numbers. A Nusselt number of 1 means that no sloshing is present and the pressure changes due to pure conduction in the stratified region (as explained in section 3.3). Only a very slight pressure change is visible. A Nusselt number of 40 results in a maximum pressure drop rate of -0.26 kPa/s and a pressure drop of 25 kPa, and a Nusselt number of 80 results in a maximum pressure drop rate of -0.49 kPa/s and a pressure drop of 43 kPa.

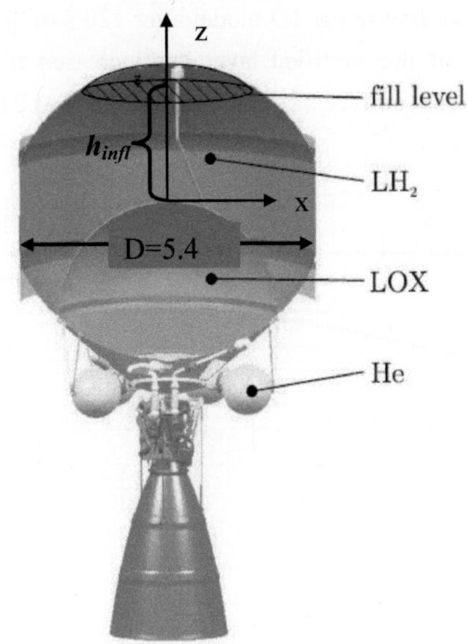

Figure 5-17. ESC-B upper stage [42] including (estimated) dimensions.

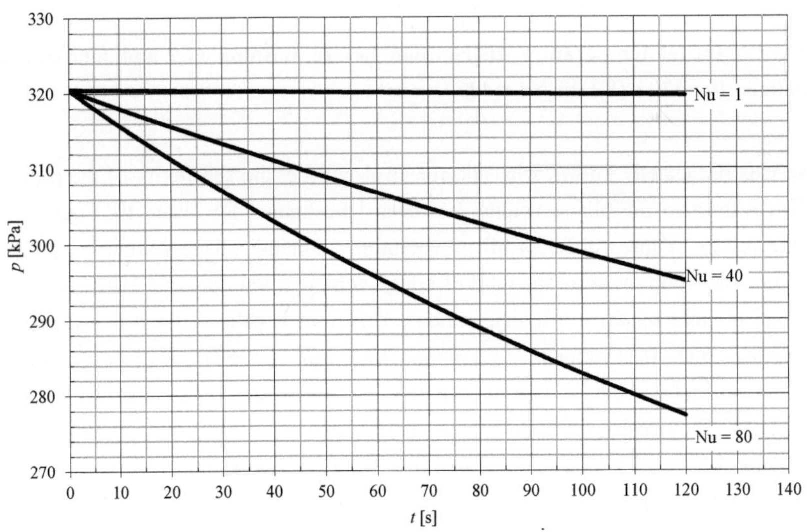

Figure 5-18. Pressure drop in ESC-B for different Nusselt numbers.

In Figure 5-19 the temperature distribution in the liquid is shown as assumed before sloshing (T_i) and according to the 1D model after 120 s of sloshing (T_f) for Nu = 40. After 120 s the thickness of the stratified layer has increased from 0.07 m to about 0.15 m. Temperature at the liquid surface has dropped from 25 K to 24.5 K.

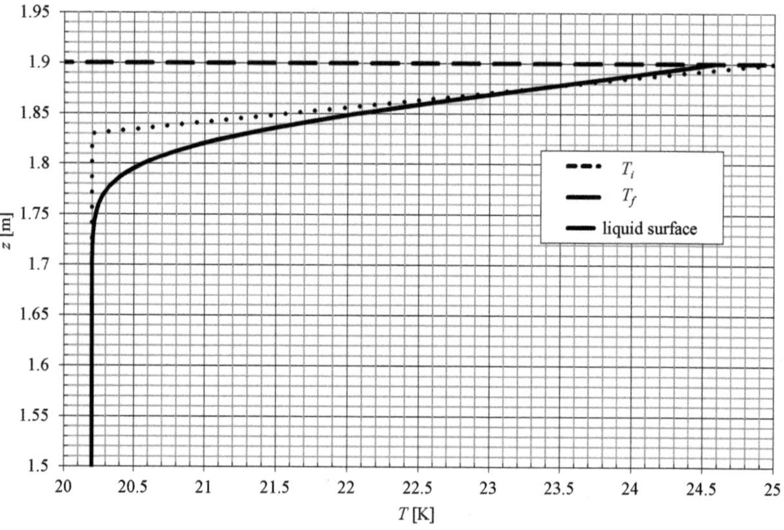

Figure 5-19. Intital temperature distribution before sloshing (T_i) and temperature distribution after 120 s (T_f) for Nu = 40.

Because it is not exactly known what the fill level will be, the sensitivity of the pressure drop with respect to the ullage volume has been investigated for Nu = 40. Figure 5-20 shows the results of three different ullage volumes. A smaller ullage volume results in a stronger pressure drop, because the ratio of ullage volume to liquid surface area is smaller (the system is more sensitive to condensation). This ratio is called L_u and is estimated at 0.49 m for $V_u = 5$ m^3, 0.31 m for $V_u = 3.2$ m^3 and 0.17 m for $V_u = 1$ m^3.

The maximum rate of pressure drop in case of $V_u = 5$ m^3 is -0.20 kPa/s, the total pressure drop is 20 kPa. The maximum rate of pressure drop in case of $V_u = 3.2$ m^3 is -0.26 kPa/s, the total pressure drop is 25 kPa. The maximum rate of pressure drop in case of $V_u = 1$ m^3 is -0.45 kPa, the total pressure drop is 40 kPa.

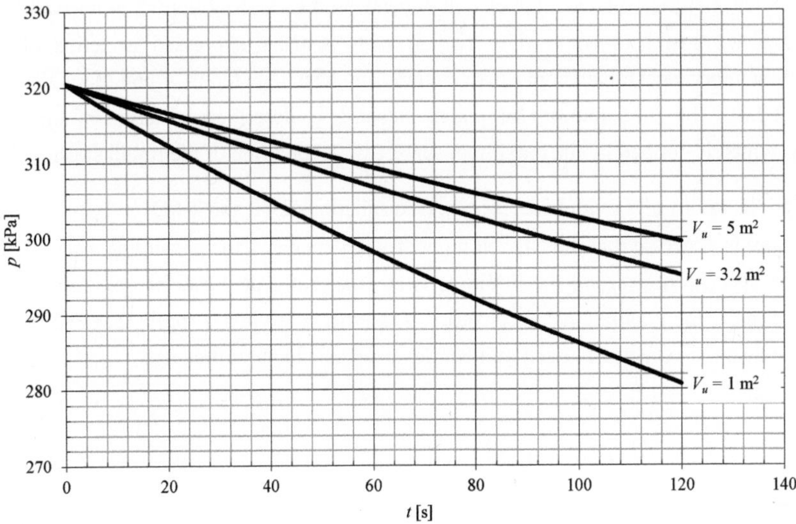

Figure 5-20. Pressure drop for different ullage volumes for Nu = 40.

5.6   Nusselt number dependency

Simulations by the 1D model depend critically on the Nusselt number. It would be very convenient if the Nusselt number could be determined without executing complicated experiments. This way a-priori knowledge of the Nusselt number combined with the 1D model would result in a simple but powerful engineering tool for the prediction of heat and mass transfer in tanks subjected to sloshing.

5.6.1   Dimensional analysis

The first step for obtaining a Nusselt number relation is to apply a dimensional analysis to see on what physical parameters the Nusselt number depends.

The analysis in section 5.2.3 has shown that the Nusselt number is independent on the pressurization duration, maximum pressure in the dewar and the thermal gradient in the liquid.

It is expected that the Nusselt number will depend on the fluid motion as is normally the case for heat transfer in convective fluid systems (usually expressed in terms of the Reynolds number, see for example the relations for heat transfer in a boundary layer in section 2.2.2). The fluid motion for planar sloshing is described in terms of the fluid velocity, which is defined by the maximum surface deflection rate of the liquid slosh motion $\dot{\Psi}_{max}$ and the radius of the liquid surface R_{ls} (see equations 2-63 and 2-65).

Furthermore is expected that the Nusselt number will depend on the density ρ_l, the viscosity μ_l and the specific heat $c_{p,l}$ of the liquid as is also normally the case for convective fluid systems. Finally the gravitational acceleration g is expected to influence the behavior of the system, and so the expected Nusselt number dependency can be written as :

$$\mathrm{Nu} = \frac{\lambda_{eff}}{\lambda} = f\left(\dot{\Psi}_{max}, R_{ls}, \rho_l, \mu_l, c_{p,l}, g\right) \qquad \text{5-27}$$

The dependency can be greatly simplified by applying a dimensional analysis in the form the Buckingham pi theorem. First of all equation 5-27 is re-written as:

$$g\left(\dot{\Psi}_{max}, R_{ls}, \rho_l, \mu_l, c_{p,l}, g, \lambda_{eff}, \lambda\right) = 0 \qquad \text{5-28}$$

The parameters defined in equation 5-28 can be expressed in the fundamental dimensions mass [kg], length [m], time [s] and temperature [K]:

$\dot{\psi}_{max}$ $[\mathrm{s}^{-1}]$

R_{ls} $[\mathrm{m}]$

ρ_l $[\mathrm{kg\ m^{-3}}]$

μ_l $[\mathrm{kg\ m^{-1}\ s^{-1}}]$

$c_{p,l}$ $[\mathrm{m^2\ K^{-1}\ s^{-2}}]$

g $[\mathrm{m\ s^{-2}}]$

λ_{eff} $[\mathrm{kg\ m\ K^{-1}\ s^{-3}}]$

λ $[\mathrm{kg\ m\ K^{-1}\ s^{-3}}]$

From the above it is seen that there are eight parameters and four fundamental dimensions, so it is expected that four dimensionless Π products are needed.

The first Π product is defined as $\Pi_1 = f_1\left(R_{ls}, \mu_l, g, \lambda, \lambda_{eff}\right)$. To ensure that Π_1 is dimensionless the relation $\Pi_1 = R_{ls}{}^0 \mu_l{}^0 g^0 \lambda^{-1} \lambda_{eff} = \dfrac{\lambda_{eff}}{\lambda}$ is obtained. The first Π product is thus:

$$\Pi_1 = Nu \qquad\qquad 5\text{-}29$$

The second Π product is defined as $\Pi_2 = f_2\left(R_{ls}, \mu_l, g, \lambda, c_{p,l}\right)$. To ensure that Π_2 is dimensionless the relation $\Pi_2 = R_{ls}{}^0 \mu_l{}^1 g^0 \lambda^{-1} c_{p,l} = \dfrac{\mu c_{p,l}}{\lambda}$ is obtained. The second Π product is thus:

$$\Pi_2 = Pr \qquad\qquad 5\text{-}30$$

The third Π product is defined as $\Pi_3 = f_3\left(R_{ls}, \mu_l, g, \lambda, \rho_l\right)$. To ensure that Π_3 is dimensionless the relation $\Pi_3 = R_{ls} \sqrt{R_{ls}} \mu_l{}^{-1} \sqrt{g} \lambda^0 \rho_l = \dfrac{\rho_l R_{ls} \sqrt{g R_{ls}}}{\mu_l}$ is obtained. The third Π product is thus a Reynolds number considering $\sqrt{g R_{ls}}$ is a characteristic velocity. However the third Π product can also be defined in terms of a Galileo number (the Galileo number is defined as $Ga = \dfrac{g L_c{}^3}{v^2}$ and expresses the ratio between the gravitational force on the fluid and the viscous force):

$$\Pi_3 = Re = \sqrt{Ga} \qquad\qquad 5\text{-}31$$

The fourth Π product is defined as $\Pi_4 = f_4\left(R_{ls}, \mu_l, g, \lambda, \dot{\Psi}_{max}\right)$. To ensure that Π_4 is dimensionless the relation $\Pi_4 = \sqrt{R_{ls}} \mu_l{}^0 \dfrac{1}{\sqrt{g}} \lambda^0 \dot{\Psi}_{max} = \dot{\Psi}_{max} \sqrt{\dfrac{R_{ls}}{g}}$ is obtained. The fourth Π product is thus:

$$\Pi_4 = \dot{\Psi}_{max} \sqrt{\frac{R_{ls}}{g}}$$

5-32

The physical relation according to equation 5-28 may thus be expressed as:

$$Nu = f_5\left(Pr, \sqrt{Ga}, \dot{\Psi}_{max} \sqrt{\frac{R_{ls}}{g}} \right)$$

5-33

5.6.2 Application to the experiments

During the experiments at ZARM the dewar geometry as well as the excitation of the dewar, the fill level and the working fluid were not changed so unfortunately the relation defined in equation 5-33 can not be defined more precise. Only the experiment conducted by Moran et al. [18] presented in section 5.4 can be used as additional data.

The experiment performed by Lacapere et al. [19] which is presented in the same section is not suitable because sloshing occurred near the eigenfrequency of the system and as such no planar slosh motion was present. Therefore equation 5-33 is not valid as the fluid motion is not described accurately anymore using $\dot{\Psi}_{max}$ and R_{ls}.

The dimensionless parameters defined in equation 5-33 are presented in Table 5-5. For the determination of dimensionless parameters for RGD 869 the data in Table 5-3 is used. The radius of the liquid surface is determined using geometrical relationships for a sphere with a radius of 0.748 m (see section 1.1) and an ullage volume of 33% of the total volume of the sphere [18]. This yields R_{ls} =0.73 m. The surface deflection rate $\dot{\Psi}_{max}$ is determined from equation 2-63 with $\sin(\omega t) = 1$, $x_a = 0.0127$ m [18], $L_1 = 0.41$ m (section 5.4), $\omega_{11} = 4.9$ rad s^{-1} (the relation for the eigenfrequency of a pendulum is used which yields $\omega_{11} = \sqrt{\frac{g}{L_1}}$) and $\beta = 2\gamma\omega_{11} = 2.5E\text{-}03$ rad s^{-1} with $\gamma = 2.6E\text{-}04$ from equation 2-60.

For the ZARM experiments $\lambda_l = 0.15$ W m^{-2}, $\rho_l = 805$ kg m^{-3} , $c_p = 2040$ J kg^{-1} K^{-1} and μ_l =160 µPa s^{-1} is used according to appendix VII. The liquid surface radius is equal to the dewar radius and is $R_{ls} = 0.145$ m (see section 3.1). The surface deflection rate $\dot{\Psi}_{max}$ is determined from equation 2-63 with $\sin(\omega t) = 1$, $x_a = 0.01$ m (see section 3.1), $L_1 = 0.079$ m

(Table 2-3), ω_{11} = 11.2 rad s^{-1} (see section 3.1) and $\beta = 2\gamma\omega_{11}$ = 1.9E-02 rad s^{-1} with γ = 8.5E-04 from equation 2-58.

	Nu	Pr	$\sqrt{Ga}$	$\dot{\Psi}_{max}\sqrt{\dfrac{R_{ls}}{g}}$
ZARM	44	2.2	8.7 E05	0.23
RDG 869 Moran et al. [18]	11	1.2	1.27 E07	0.15

Table 5-5. Dimensionless parameters for the experiments of ZARM and RDG 869 from Moran et al. [18].

Unfortunately two data sets do not suffice to determine an empirical relation between the dimensionless parameters. Future experiments should be performed to gather more datasets. The slosh motion should be varied as well as the fluid type and the dewar geometry such that the influence of all relevant dimensionless parameters on the Nusselt number can be studied.

6   Summary

The topic of this work was to develop simple engineering models for the simulation of cryogenic sloshing taking into account heat and mass transfer. Such a model could then be used to make predictions regarding to pressure drops in full scale launch vehicle tanks containing cryogenic propellants.

The results of experiments conducted at ZARM by Arndt [42] are used as a reference for the development of the model. These experiments were executed in a cylindrical glass dewar with a radius of $R = 0.145$ m and a spherical bottom. The test fluid was LN_2 which has been used as a replacement for the cryogenic liquids LH_2 and LOX usually present in cryogenic launcher stages.

The dewar was closed and pressurised to different pressure levels by means of two different pressurisation methods, namely active pressurisation and self-pressurisation. In case of active pressurisation, gaseous nitrogen was actively added to the ullage volume increasing the pressure in the dewar. Self-pressurisation means that pressure was increased by unavoidable heat leaks entering the dewar from the surroundings. This caused some of the LN_2 to evaporate and pressure to increase. Self-pressurisation involves much longer times scales of up to an hour compared to active pressurisation where times scales of a maximum of two minutes were present.

After the pressure had reached the desired level the sloshing was initiated. A sloshing motion with a frequency of $\omega = 0.78\ \omega_{11}$ is introduced with an excitation amplitude of $x_a = 0.07\ R$. This resulted in a well-defined planar slosh motion without swirling. After initiation of the sloshing the pressure in the system dropped to a certain minimum level, where after pressure started to rise slowly again.

Experimental results show a strong thermal gradient in the liquid near the liquid surface after the pressurisation. During sloshing, the gradient is destructed and the temperature of the liquid near the liquid surface decreases. Analyses of the experimental results indicate that the pressure drop is mainly caused by condensation of the ullage vapour, caused by a decrease of the liquid temperature at the liquid surface yielding a different saturation pressure. This means that to understand and model the pressure drop, it is also necessary to be able to understand and model the formation of thermal stratification.

The CFD tool FLOW3D was used to model the pressurisation phase to help understanding the formation of thermal stratification. It turned out that by using FLOW 3D it is not possible to get results which are quantitavily correct, because the required mesh resolution is too high and yields unacceptably long computation times of several months. However, FLOW3D results can be used to investigate the heat flow mechanisms responsible for the formation of thermal stratification

The FLOW 3D results show that in case of self-pressurisation the sharp thermal gradient at the liquid surface is mainly caused by heat entering the liquid in tangential direction through the dewar wall. This is caused by the fact that the dewar wall is hotter in the ullage region und thus heat is conducted towards the liquid region. This heat is releases just below the liquid surface causing the sharp thermal gradient.

In case of active pressurisation the sharp thermal gradient is caused by the fact that actively increasing the pressure causes ullage vapour to condensate at the liquid surface (because the actively increased ullage pressure is higher that the saturation pressure belonging to the liquid temperature at the surface). The heat released by condensation heats up the liquid temperature at the surface creating the thermal gradient in the liquid.

The thickness of the thermally stratified layer increases over time as heat is conducted through the thermal layer into the bulk of the liquid. The FLOW3D results hint to the fact that a simple 1D conduction model could suffice in describing thermal stratification. A thermal conduction model has been set up resolving the wall and the liquid by using a thickness of 1cell for the liquid and 1 cell for the wall. Heat transfer in the wall and liquid and between the wall and liquid is assumed to take place by conduction only. The model is strictly speaking a 2D model but because there are only 2 cells in the x direction and a few hundred in the z direction the model is referred to as a 1D model. The stratification predicted by the model shows close agreement to the experimentally measured stratification.

After having successfully developed a model for the stratification, the next step was to try to model the pressure drop caused by sloshing. It is shown that FLOW3D cannot be used to simulate the pressure drop caused by liquid slosh due to a limitation of the numerical method. But by setting up a 1D energy equation for liquid at the liquid surface during sloshing it is shown that the 1D model used for the determination of thermal stratification can also be used to describe the pressure drop during sloshing. To this end, the liquid heat conduction coefficient must be increased in liquid regions influenced by the sloshing motion. This increase of heat transfer within the liquid is explained by convective motions resulting from the slosh motion.

The increase of the heat conduction coefficient is modelled by a Nusselt number. This Nusselt number can be determined from the experiments by using the experimentally measured rate of pressure drop. Using this, the mass flow rate of condensated ullage vapour can be determined which in turn yields the energy transported over the liquid surface. Insertion of this Nusselt number into the 1D model yields pressure drop curves comparable with those measured in the experiments. Also the temperature development in the liquid during sloshing according to the 1D model is in agreement with the experimental data.

The 1D slosh model is then applied to experiments conducted by Moran et al. [18] and Lacapere et al. [19]. The pressure development during sloshing determined with the model is in close agreement to the experimental results. This is followed by applying the 1D model to predict the pressure development during sloshing of the cryogenic ESC-B upper stage currently under development at Astrium. To this end some assumptions have to be made regarding initial stratification in the liquid and on the Nusselt number.

Finally a dimensional analysis is performed to determine the relevant dimensionless parameters which define the Nusselt number. This can be used in future work to define an empirical relation to determine the Nusselt number a-priori as an input for the 1D model.

7 Conclusions & recommendations

7.1 Conclusions

The following conclusions can be drawn from this work:

- The pressure drop observed in the experiments is mainly caused by condensation
- The condensation process is explained by the destruction of the thermal gradient in the liquid near the liquid surface, causing the temperature at the liquid surface to drop
- The sharp thermal gradient is mainly caused by heat transported in tangential direction through the dewar wall from the hotter ullage region into the liquid
- The stratification process can be modelled efficiently by a 1D conduction model
- The pressure drop during sloshing can be modelled by using the same 1D approach but by increasing the heat conduction coefficient in the liquid
- The increase of the heat conduction coefficient is described by a Nusselt number, which can be obtained from the experimental results

7.2 Recommendations

The 1D engineering approach using an experimentally determined Nusselt number is able to reproduce the experimental results. This gives confidence in this approach, and some suggestions for future research can be made to improve it.

The 1D model would be more powerful if the Nusselt number could be estimated a-priori without doing experiments. More experiments should be executed changing the excitation frequency and amplitude to create different sloshing motions. The dependency of the Nusselt number on the sloshing motion can then be investigated. Also, other fluids should be used, for example LOX or LH_2, to check the dependency on Prandtl number and viscosity. Finally, it is recommended to perform additional experiments using a dewar with other dimensions to investigate the dependency on the Galileo number.

8 References

[1] Ariane 5 user manual, Arianespace, Issue 5 Revision 0, July 2008

[2] Sewall, J.L.: An Experimental and Theoretical Study of the Effect of Fuel on Pitching-Translation Flutter, NACA TN 4166, Dec. 1957

[3] Goldsborough, R .: The Tidal Oscillations in an Elliptic Basin of Variable Depth, Proc. Roy. Soc. (London), A, 130, pp. 157-167, 1930

[4] Werner, P.W.; and Sundquist, J.: On Hydrodynamic Earthquake Effects, Trans. Am. Geophysical Union, vol. 30, no. 5, pp. 636-657, Oct. 1949

[5] Abramson, H. N., "The Dynamic Behaviour of Liquids in Moving Containers" Tech. Report NASA-SP-106, 1966

[6] Dodge, F. T., "The new dynamic behaviour of liquids in moving containers", SwRI San Antonio, Texas, 2000

[7] Miles, J.W., "Resonantly forced surface waves in a circular cylinder", J. Fluid Mech., 149, pp. 15-31, 1984

[8] Ring, E., Ezra, A.A., Bowman, J.A., Thorson, M.H., Brown, C.D., Morgan, S.K., Lukens, S.C.: Rocket Propellant and Pressurization Systems, Prentice-Hall, Inc, Englewood Cliffs, N.J., 1964

[9] Clark, J.A.: A review of pressurization, stratification and interfacial phenomena, University of Michigan, 1965

[10] Grayson, G., Lopez, A., Chandler, F.: Cryogenic tank modelling for the Saturn AS-203 experiment, 42nd AIAA Joint Propulsion Conference, AIAA-2006-5258, Sacramento, CA, USA, 2006

[11]   Schallhorn, P., Grob, L.: Upper Stage Tank Thermodynamic Modeling Using SINDA/FLUINT, 42nd AIAA Joint Propulsion Conference, AIAA-2006-5051, Sacramento, CA, USA, 2006

[12]   Oliveira, J., Kirk, D.R., Schallhorn, P.A.: Cryogenic Propellant Stratification in a Rotating and Reduced Gravity Environment, 43rd AIAA Joint Propulsion Conference, AIAA 2007-5495, Cincinnati, OH, 2007

[13]   Ibrahim, R.A., Liquid Sloshing Dynamics, Cambridge University Press, 2005

[14]   Bauer, H.: Hydroelastische Schwingungen im aufrechten Kreiszylinderbehälter, Z. Flugwiss., 18, pp. 117-134, 1970

[15]   Mikishev, G.N., Dorozhkin, N.Y., An experimental investigation of free oscillations of a liquid in containers, Izv. Akad., Nauk SSSR, Otd. Tekh., Nauk, Mekh. I Mashinor, 4, pp. 48-83, 1961

[16]   Stephens, D.G., Leonard, H.W., Silveira, M.A., Investigation of the damping of liquids in right-circular cylindrical tanks, NASA TN D-1367, 1962

[17]   Boyce, W.E., DiPrima, R.C.: Elementary differential equations and boundary value problems, 6th edition, John Wiley & sons, 1997

[18]   Moran, E.M., McNeils, N.B., Kudlac, M.T., Haberbusch, M.S., Satornino, G.A.: Experimental results of hydrogen slosh in a 60 cubic foot (1750 liter) tank. 30th Joint Propulsion Conference, AIAA 94-3259, 1994

[19]   Lacapere, J., Vieille, B., Legrand, B.: Experimental and numerical results of sloshing with cryogenic fluids, 2nd European Conference for Aerospace Sciences (EUCASS) 2007

[20]   Das, S. P., Hopfinger, E. J.: Mass transfer enhancement by gravity waves at a liquid vapour interface, Int. J. Heat Mass Tran., Vol. 52, No. 5-6, pp. 1400-1411, 2009

[21]   Chin, J.H. et al..: Analytical and experimental study of liquid orientation and stratification in standard and reduced gravity fields, Lockheed missiles & space company, Rept. 2-05-64-1, July 1964

[22]   Anderson, J.D. Jr., Fundamentals of aerodynamics, McGraw-Hill, 1991

[23]   Cengel, Y.A.: Heat transfer: a practical approach, McGraw Hill Education, USA, 2002

[24]   Churchill, S.W., Chu, H.H.S.: Correlating equations for laminar and turbulent free convection from a vertical plate, Int. J. Heat Mass Transfer 18, pp. 1323-1329, 1975

[25]   Le Fèvre, E.J.: Laminar free convection from a vertical plane surface. Proc. 9th Int. Congr. Appl. Mech., Brussels, paper I-168, 1956

[26]   McAdams, Wm. H.: Heat transmission, 3rd edition New York, Mc Graw-Hill book company, 1954

[27]   Carey, Van P.: Liquid-Vapor Phase-Change Phenomena: An Introduction to the Thermophysics of Vaporization and Condensation Processes in Heat Transfer Equipment, Hemisphere Publishing Cooperation, 1992

[28]   Dreyer, M.E., Strömungen mit freien Oberflächen, Fachgebiet Technische Mechanik, Strömungslehre, Raumfahrttechnology und Mikrogravitation, Universität Bremen, 2010

[29]   Baehr, H.D., Stephan, K.: Wärme- und Stoffübertragung, Springer-Verlag, 1994

[30]   Isashenko, V.P., Condensation heat transfer (Russian), Moscow, Energia, 1977

[31]   Colburn, A.P.: The calculation of condensation when a portion of the condensate layer is in turbulent flow, Trans. AIChE, vol. 30, pp. 187-193, 1933

[32]   Grober, H., Erk, S., Grigull, U.: Fundamentals of heat transfer, McGraw-Hill, New York, 1961

[33] Arndt, T., Dreyer, M., Behruzi, P., Winter, M., Van Foreest, A.: Cryogenic Sloshing Tests in a Pressurized Cylindrical Reservoir, Proceeding at the 45th AIAA Joint Propulsion Conference, AIAA 2009-4860, 2009

[34] Document received from FLOW Science in private communication: Representation of Internal Energy and Temperature in the Two-Fluid Model in FLOW-3D[®], received 26˜feb-2009

[35] Van Foreest, A., Dreyer, M., Arndt, T.: Moving Two-Fluid Systems Using the Volume-of-Fluid Method and Single-Temperature Approximation, AIAA Journal, vol.49 no.12 ,2011

[36] Moran, M.J., Shapiro, H.N.: Fundamentals of engineering thermodynamics, John Wiley & sons, England, 1998

[37] Baehr, H.D.: Thermodynamik, Springer, Germany, 1996

[38] Van Kimmenaede, A.J.M.: Warmteleer voor technici, Educaboek, The Netherlands, 1990

[39] Hopfinger, E. J., Das, S. P.: Mass transfer enhancement by capillary waves at a liquid vapour interface, Exp. Fluids., Volume 46, Number 4, 597-605, 2008

[40] Olson, J.R., Fischer H.E., Pohl, R.O., Effect of Crystallization on Thermal Conductivity and Specific Heat of Two Corning Glass-Ceramics, J. Am. Ceram. Soc., 1991

[41] FLOW 3D User Manual version 9.4, Flow Science, Inc., 2009

[42] Arndt, T: Sloshing of cryogenic liquids under normal gravity conditions, PhD thesis, University of Bremen, 2011

I. CAD drawing of the dewar

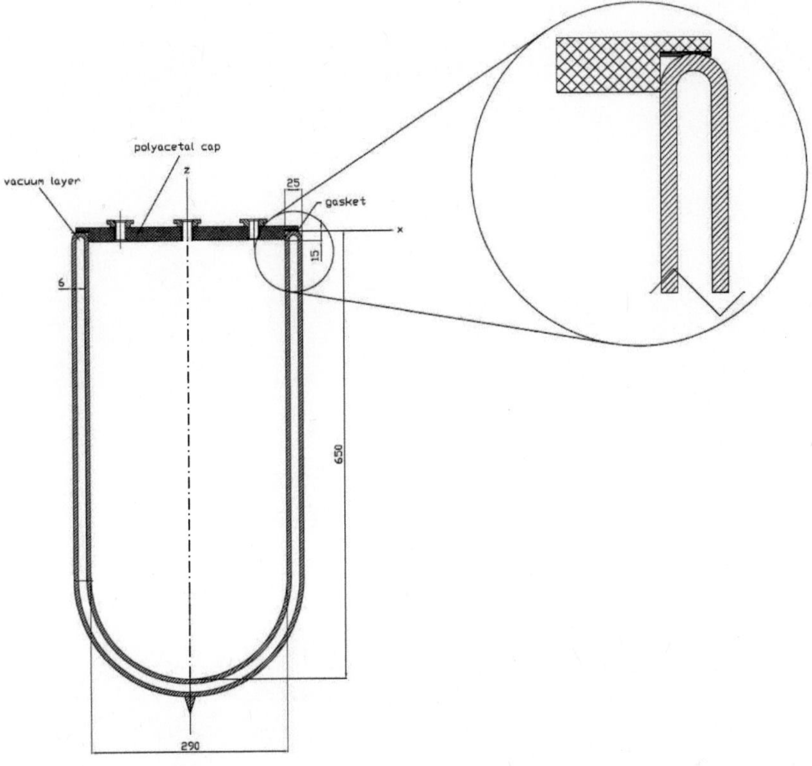

II. Bessel Function

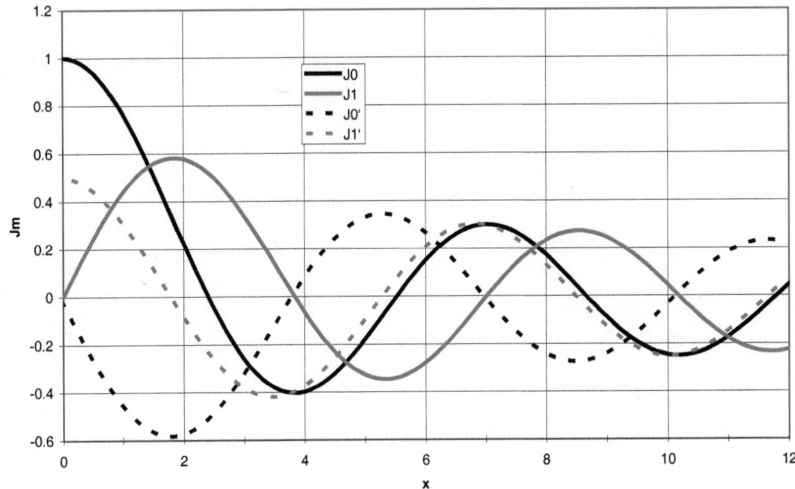

Figure II-1. Zero and first order Bessel functions and their derivatives.

III. Experimental data

Exp	p_{max} [kPa]	t_{press} [s]	$\Delta T_l\|_{ts}$ [K]	$\delta_T\|_{ts}$ [m]	T_v^{av} [K]	$m_v\|_{tb}$ [kg]	$m_v\|_{ts}$ [kg]	m_{evap} [kg]
sp120	120 ±0.1%	946	1.08 ±0.031	0.023 ±10%	167	0.048 ±2.8E-04	0.055 ±3.2E-04	0.0066 ±4.2E-04
sp130	130 ±0.1%	1408	1.70 ±0.031	0.028 ±10%	169	0.048 ±2.8E-04	0.058 ±3.4E-04	0.0109 ±4.4E-04
sp140	140 ±0.1%	2143	2.26 ±0.031	0.035 ±10%	168.2	0.049 ±2.8E-04	0.063 ±3.7E-04	0.0145 ±4.6E-04
sp160	160 ±0.1%	3720	3.38 ±0.031	0.046 ±10%	168.5	0.049 ±2.8E-04	0.072 ±4.2E-04	0.0228 ±5.1E-04
ap120	120 ±0.1%	38	1.20 ±0.031	0.0047 ±10%	170.6	0.047 ±2.8E-04	0.053 ±3.1E-04	
ap130	130 ±0.1%	67	1.70 ±0.031	0.0062 ±10%	168.6	0.048 ±2.8E-04	0.059 ±3.4E-04	
ap140	140 ±0.1%	95	2.36 ±0.031	0.0074 ±10%	167.5	0.049 ±2.8E-04	0.063 ±3.7E-04	
ap150	150 ±0.1%	119	2.89 ±0.031	0.0082 ±10%	168	0.049 ±2.8E-04	0.068 ±3.9E-04	
ap160	160 ±0.1%	141	3.43 ±0.031	0.0090 ±10%	167.7	0.049 ±2.8E-04	0.072 ±4.2E-04	

Table III-1. Characteristic experimental data for the pressurisation phase

Exp	p_{max}	p_{min}	t_{pmin}	p_{tpd}	$-\left.\left\|\dfrac{dp}{dt}\right\|\right._{max}$	t_{pd}	$\Delta T_l\|_{tpd}$	$\delta_T\|_{tpd}$	h_e	$T_{v,av}\|_{tpd}$	$m_v\|_{tpd}$	$\dot{m}_{pch}\|_{tpd}$	$\dot{Q}_{conv}\|_{tpd}$	$\lambda_{l,eff}$	Nu
	[kPa]	[kPa]	[s]	[kPa]	[kPa s^{-1}]	[s]	[K]	[m]	[kJ kg^{-1}]	[K]	[kg]	[kg s^{-1}]	[W]	[W m^{-1} K^{-1}]	[-]
sp120	120 ±0.1%	114.2 ±0.1%	220.1	119.5 ±0.1%	-0.11 ±5%	36.7	1.04 ±0.031	0.039 ±10%	-119.01	164.5	0.055 ±3.1E-4	-5.1E-05 ±2.5E-06	-12 ±1.34	6.88 ±0.8	46 ±5.4
sp130	130 ±0.1%	119.5 ±0.1%	327.9	129.6 ±0.1%	-0.16 ±5%	27.9	1.68 ±0.031	0.038 ±10%	-117.54	165.4	0.060 ±3.4E-04	-7.3E-05 ±3.6E-06	-18 ±1.50	6.05 ±0.6	40 ±3.7
sp140	140 ±0.1%	124.9 ±0.1%	406.4	137.9 ±0.1%	-0.20 ±5%	30.7	2.13 ±0.031	0.045 ±10%	-116.15	166.4	0.063 ±3.6E-04	-9.1E-05 ±4.5E-06	-22 ±1.64	7.03 ±0.6	47 ±3.8
sp160	160 ±0.1%	134.5 ±0.1%	688.3	157.9 ±0.1%	-0.25 ±5%	26	3.26 ±0.031	0.053 ±10%	-113.57	167.3	0.072 ±4.1E-04	-1.1E-04 ±5.7E-06	-27 ±1.90	6.59 ±0.5	44 ±3.3
ap120	120 ±0.1%	112.7 ±0.1%	173.6	120.0 ±0.1%	-0.13 ±5%	10	1.20 ±0.031	0.007 ±10%	-119.01	163	0.056 ±3.1E-04	-1.4E-05 ±2.9E-06	-9 ±1.38	0.86 ±0.1	6 ±0.9
ap130	130 ±0.1%	114.6 ±0.1%	218.2	127.0 ±0.1%	-0.38 ±5%	10	1.50 ±0.031	0.017 ±10%	-117.54	165.5	0.058 ±3.3E-04	-1.4E-04 ±8.6E-06	-37 ±1.99	6.30 ±0.4	42 ±2.6
ap140	140 ±0.1%	117.1 ±0.1%	311.6	134.9 ±0.1%	-0.51 ±5%	10	2.03 ±0.031	0.018 ±10%	-116.15	165.8	0.062 ±3.5E-04	-2.0E-04 ±1.2E-05	-52 ±2.43	6.79 ±0.4	45 ±2.6
ap150	150 ±0.1%	119.7 ±0.1%	339.3	147.4 ±0.1%	-0.85 ±5%	5	2.73 ±0.031	0.014 ±10%	-114.83	166.6	0.067 ±3.9E-04	-3.4E-04 ±1.9E-05	-87 ±3.62	6.75 ±0.3	45 ±2.3
ap160	160 ±0.1%	121.9 ±0.1%	487.4	157.7 ±0.1%	-1.0 ±5%	5	3.30 ±0.031	0.015 ±10%	-113.57	166.5	0.072 ±4.1E-04	-4.1E-04 ±2.3E-05	-103 ±4.17	6.88 ±0.3	46 ±2.3

Table III-2. Characteristic experimental data for the sloshing phase

	z [m]	T [K] @ t_b	T [K] @ t_s	T [K] @ t_{pmin}
T15	0.003	77.58	77.73	77.73
T14	0.103	77.66	77.72	77.73
T13	0.203	77.6	77.72	78.12
T12	0.253	77.58	77.79	78.25
T11	0.278	77.58	78.4	78.31
T$_{sat}$	0.29	77.36	78.79	78.32
T4	0.334	106.73	112.87	102.1
T8	0.334	108	114.1	103.62
T3	0.384	128.9	134.93	134.74
T7	0.384	129.23	135.22	134.9
T2	0.434	156.67	161.73	162.88
T6	0.434	157.44	162.47	163.85
T1	0.484	185.93	190.11	191.75
T5	0.484	187.31	191.41	192.9
T9	0.55	265	259.33	258.3
T10	0.635	285.77	285.45	285.3

Table III-3. Temperature data for sp120

	z [m]	T [K] @ t_b	T [K] @ t_s	T [K] @ t_{pmin}
T15	0.003	77.55	77.69	77.73
T14	0.103	77.57	77.77	77.75
T13	0.203	77.66	77.85	78.04
T12	0.253	77.53	77.81	78.63
T11	0.278	77.48	78.38	78.7
T$_{sat}$	0.29	77.41	79.5	78.76
T4	0.334	107.61	114.45	100.83
T8	0.334	108.99	115.68	102.59
T3	0.384	129.95	137.06	136.24
T7	0.384	130.32	137.24	136.74
T2	0.434	155.82	163.08	164.63
T6	0.434	156.48	163.74	165.44
T1	0.484	184.52	192.13	193.54
T5	0.484	185.56	193.47	194.89
T9	0.55	290.12	267.71	265.41
T10	0.635	288.17	287.7	287.64

Table III-4. Temperature data for sp130

	z [m]	T [K] @ t_b	T [K] @ t_s	T [K] @ t_{pmin}
T15	0.003	77.61	77.83	77.86
T14	0.103	77.64	77.85	77.88
T13	0.203	77.74	78	78.31
T12	0.253	77.6	78.04	79.03
T11	0.278	77.59	79.07	79.07
T$_{sat}$	0.29	77.41	80.16	79.15
T4	0.334	104.43	115.74	102.92
T8	0.334	105.66	116.92	104.07
T3	0.384	126.67	138.22	137.53
T7	0.384	126.92	138.4	137.57
T2	0.434	155.24	163.56	164.75
T6	0.434	155.94	164.18	165.36
T1	0.484	183.65	189.44	190.37
T5	0.484	184.82	190.54	191.46
T9	0.55	253.63	250.97	250.67
T10	0.635	286.69	286.04	285.89

Table III-5. Temperature data for sp140

	z [m]	T [K] @ t_b	T [K] @ t_s	T [K] @ t_{pmin}
T15	0.003	77.55	77.93	78.02
T14	0.103	77.58	77.95	77.88
T13	0.203	77.59	78	79.18
T12	0.253	77.55	78.54	79.71
T11	0.278	77.55	80.51	79.78
T_{sat}	0.29	77.37	81.39	79.81
T4	0.334	105.56	118.87	104.95
T8	0.334	106.68	119.92	106.05
T3	0.384	125.64	140.32	139.03
T7	0.384	125.87	140.47	138.97
T2	0.434	151.42	163.59	164.89
T6	0.434	152.02	164.12	165.39
T1	0.484	177.74	187.39	188.65
T5	0.484	178.8	188.36	189.69
T9	0.55	242.99	245.43	245.72
T10	0.635	284.42	284.37	284.41

Table III-6. Temperature data for sp160

	z [m]	T [K] @ t_b	T [K] @ t_s	T [K] @ t_{pmin}
T15	0.003	77.63	77.58	77.61
T14	0.103	77.62	77.59	77.63
T13	0.203	77.61	77.6	77.67
T12	0.253	77.59	77.58	78.06
T11	0.278	77.57	77.61	78.15
T_{sat}	0.29	77.41	78.78	78.26
T4	0.334	105.71	122.67	107.39
T8	0.334	107.59	123.34	108.17
T3	0.384	141.19	146.68	141.89
T7	0.384	140.81	146.18	141.56
T2	0.434	164.44	168.66	164.65
T6	0.434	165.42	168.61	165.07
T1	0.484	185.58	189.76	185.97
T5	0.484	186.59	190.63	186.51
T9	0.55	235.08	236.42	235.26
T10	0.635	279.72	279.75	279.72

Table III-7. Temperature data for ap120

	z [m]	T [K] @ t_b	T [K] @ t_s	T [K] @ t_{pmin}
T15	0.003	77.75	77.78	77.75
T14	0.103	77.76	77.77	77.78
T13	0.203	77.85	77.86	78.03
T12	0.253	77.7	77.75	78.26
T11	0.278	77.7	77.77	78.32
T_{sat}	0.29	77.41	79.5	78.4
T4	0.334	110.23	117.25	103.34
T8	0.334	111.09	117.88	104.84
T3	0.384	133.61	142.13	137.26
T7	0.384	133.38	141.6	136.26
T2	0.434	159.48	168.38	163.86
T6	0.434	159.66	168.45	164.4
T1	0.484	182.57	191.54	186.8
T5	0.484	183.19	192.54	187.05
T9	0.55	233.33	236.71	234.97
T10	0.635	279.35	279.36	279.35

Table III-8. Temperature data for ap130

	z [m]	T [K] @ t_b	T [K] @ t_s	T [K] @ t_{pmin}
T15	0.003	77.76	77.79	77.83
T14	0.103	77.79	77.8	77.87
T13	0.203	77.9	77.91	78.28
T12	0.253	77.75	77.82	78.47
T11	0.278	77.75	77.9	78.51
T$_{sat}$	0.29	77.41	80.16	78.59
T4	0.334	106.78	115.46	102.74
T8	0.334	107.65	116.14	103.93
T3	0.384	129.38	139.87	135.71
T7	0.384	129.12	139.38	135.43
T2	0.434	156	167.09	163.34
T6	0.434	156.14	167.22	163.37
T1	0.484	180.41	191.44	185.76
T5	0.484	181.08	192.42	187.18
T9	0.55	233.46	237.8	235.37
T10	0.635	279.06	279.11	279.12

Table III-9. Temperature data for ap140

	z [m]	T [K] @ t_b	T [K] @ t_s	T [K] @ t_{pmin}
T15	0.003	77.85	77.89	77.91
T14	0.103	77.87	77.91	77.95
T13	0.203	77.94	78	78.49
T12	0.253	77.79	77.89	78.67
T11	0.278	77.8	78.11	78.73
T$_{sat}$	0.29	77.41	80.79	78.78
T4	0.334	107.26	116.89	104.12
T8	0.334	108.03	117.48	104.76
T3	0.384	128.45	139.94	135.25
T7	0.384	128.25	139.52	134.67
T2	0.434	154.25	166.66	161.83
T6	0.434	154.43	166.85	162.18
T1	0.484	179.12	191.66	186.57
T5	0.484	179.71	192.58	187.12
T9	0.55	235.46	240.01	236.97
T10	0.635	278.85	278.96	279.04

Table III-10. Temperature data for ap150

	z [m]	T [K] @ t_b	T [K] @ t_s	T [K] @ t_{pmin}
T15	0.003	77.89	77.94	77.99
T14	0.103	77.95	77.97	78.02
T13	0.203	77.88	77.95	78.66
T12	0.253	77.83	77.95	78.84
T11	0.278	77.79	78.53	78.92
T$_{sat}$	0.29	77.42	81.35	78.94
T4	0.334	105.84	115.87	104.05
T8	0.334	106.6	116.47	104.74
T3	0.384	126.93	138.98	134.85
T7	0.384	126.75	138.62	134.55
T2	0.434	153.17	166.1	161.66
T6	0.434	153.38	166.32	162
T1	0.484	178.34	191.5	186.63
T5	0.484	179	192.36	187.17
T9	0.55	234.52	240.1	237.12
T10	0.635	279.03	279.13	279.24

Table III-11. Temperature data for ap160

Exp	p_{max}	t_{press}	$-\left|\dfrac{dp}{dt}\right|_{max}$	ΔT_l	δ_T	T_v	Nu	V_u	R_{ls}	L_u
	[kPa]	[s]	[kPa/s]	[K]	[m]	[K]	[-]	[m³]	[m]	[m]
ESCB1	320	3600	-0.008	4.8	0.07	31.2	1	3.2	1.8	0.31
ESCB2	320	3600	-0.26	4.8	0.07	31.2	40	3.2	1.8	0.31
ESCB3	320	3600	-0.49	4.8	0.07	31.2	80	3.2	1.8	0.31
ESCB4	320	3600	-0.45	4.8	0.07	31.2	40	1	1.37	0.17
ESCB6	320	3600	-0.2	4.8	0.07	31.2	40	5	2	0.49

Table III-12. ESC-B model data

IV. Note on c_p and c_v in liquids

In many thermodynamic textbooks, it is stated that liquids can assumed to be incompressible and therefore c_p and c_v are equal [37] [38] [23]. This is incorrect. Most liquids do have limited compressibility (i.e $v \approx$ constant) but this does not mean that c_p and c_v are equal.

The difference between the specific heats is especially large in case of cryogenic liquids. In case of for example liquid nitrogen, the difference between the specific heats is about a factor 1.8 to 1.9. This can immediately be seen by looking at the NIST property tables in appendix VII.

Values of specific heats for different liquids are compared in Table IV-1. As can be seen even for water there is still a difference of a factor 1.12.

liquid	T [K]	c_p [J kg^{-1} K^{-1}] from NIST	c_v [J kg^{-1} K^{-1}] from NIST	c_p/c_v	c_p-c_v [J kg^{-1} K^{-1}]
Nitrogen	77	2039.8	1086.1	1.88	953.7
Oxygen	90	1698.9	929.6	1.83	769.3
Hydrogen	20	9565.1	5636.8	1.70	3928.3
Water	372	4214.4	3774.1	1.12	440.3
Benzene	352	1885.4	1348.8	1.40	536.6

Table IV-1. Values of c_p and c_v at $p = 100$ kPa for different liquids according to NIST property tables.

This difference can be explained by looking at the equations of state. The specific enthalpy of a fluid is a function of temperature and pressure, $h = h(T, p)$. The differential is given by:

$$dh = \left(\frac{\partial h}{\partial T}\right)_p dT + \left(\frac{\partial h}{\partial p}\right)_T dp \qquad\qquad \text{IV-1}$$

By using the Gibbs-function it can be shown that [37]:

$$\left(\frac{\partial h}{\partial p}\right)_T = v - T\left(\frac{\partial v}{\partial T}\right)_p$$

The term $\left(\dfrac{\partial h}{\partial T}\right)_p$ is defined as the specific heat at constant pressure $c_p(T)$. Insertion into equation IV-1 leads to:

$$dh = c_p(T,p)dT + \left[v - T\left(\frac{\partial v}{\partial T}\right)_p\right]dp \qquad \text{IV-2}$$

The specific internal energy of a fluid is a function of temperature and specific volume, $i = i(T,v)$. The differential is given by:

$$di = \left(\frac{\partial i}{\partial T}\right)_v dT + \left(\frac{\partial i}{\partial v}\right)_T dv \qquad \text{IV-3}$$

By using the Helmholtz-function [37] it can be shown that:

$$\left(\frac{\partial i}{\partial v}\right)_T = -p + T\left(\frac{\partial p}{\partial T}\right)_v$$

The term $\left(\dfrac{\partial i}{\partial T}\right)_v$ is defined as the specific heat at constant temperature $c_v(T)$. Insertion into equation IV-3 leads to:

$$di = c_v(T,v)dT + \left[-p + T\left(\frac{\partial p}{\partial T}\right)_v\right]dv \qquad \text{IV-4}$$

The enthalpy of a fluid is defined as $h = i + pv$. The differential is given by:

$$dh = di + pdv + vdp \qquad \text{IV-5}$$

Inserting equation IV-4 in equation IV-5 yields:

$$dh = c_v(T,v)dT + vdp + T\left(\frac{\partial p}{\partial T}\right)_v dv \qquad \text{IV-6}$$

Combining equations IV-2 and IV-6 gives:

$$c_v(T,v) = c_p(T,p) - \frac{T}{dT}\left(\left(\frac{\partial v}{\partial T}\right)_p dp + \left(\frac{\partial p}{\partial T}\right)_v dv\right) \qquad \text{IV-7}$$

From equation IV-7 it can be seen that if v = constant, which is the case for fully incompressible fluids, the specific heats are equal. In reality v will never be completely constant, although the change might be very small. This will not always lead to the conclusion that the specific heats will be equal, because the products $\left(\frac{\partial v}{\partial T}\right)_p dp$ and $\left(\frac{\partial p}{\partial T}\right)_v dv$ still can be substantial. This can be the case for high pressure difference or a high ratio of $\left(\frac{\partial p}{\partial T}\right)_v$. It is exactly this point what is forgotten in literature.

The discussion given above is illustrated using liquid nitrogen as an example. Pressure is assumed to be constant at 100 kPa. The term $\left(\frac{\partial v}{\partial T}\right)_p dp$ is thus zero and equation IV-7 reduces to:

$$c_v(T,v) = c_p(T,p) - T\left(\frac{\partial p}{\partial T}\right)_v \frac{dv}{dT} \qquad \text{IV-8}$$

The term $\left(\frac{\partial p}{\partial T}\right)_v$ is provided by NIST property tables as well as the term $\frac{dv}{dT}$. The results are listed in Table IV-2.

What is seen that although the liquids have only very limited compressibility ($\frac{dv}{dT}$ is very small) the term $T\left(\frac{\partial p}{\partial T}\right)_v \frac{dv}{dT}$ is substantial, explaining the big difference between c_p and c_v (compare with Table IV-1).

liquid	T [K]	v [m^3 kg^{-1}]	p [kPa]	$\left(\dfrac{\partial p}{\partial T}\right)_v$ [Pa K^{-1}]	$\dfrac{dv}{dT}$ [m^3 kg^{-1} K^{-1}]	$T\left(\dfrac{\partial p}{\partial T}\right)_v\dfrac{dv}{dT}$ [J kg^{-1} K^{-1}]
Nitrogen	77	0.001238082	100	1.77E+06	7.0E-06	954
Oxygen	90	0.000875576	100	2.24E+06	3.80E-06	766
Hydrogen	20	0.01406097	100	8.45E+05	2.32E-04	3921
Water	372	0.0010426	100	1.53E+06	8.00E-07	455
Benzene	352	0.0012272	100	8.94E+05	1.70E-06	535

Table IV-2. Values according to NIST for explaing the difference between c_p and c_v.

Another commonly made mistake is that in liquids, the change in internal energy is determined by $di = c_v dT$ (i.e. the specific heat at constant volume is used and other terms are neglected). If the temperature increase of a liquid under constant pressure is considered, it is seen that by combining equation IV-4 and IV-8 the relation $di = c_p dT - p dv$ is obtained. The term $p dv$ is usually small compared to the term $c_p dT$, so that for most liquids the change in internal energy is described by using the specific heat at constant pressure and neglecting the other term:

$$di = c_p(T,v)dT$$

IV-9

This fact is easily verified by looking at the NIST property tables

V. Fitting of the pressure curve

The pressure curves are fitted with a polynomial of the nth order in the form of:

$$p(x) = p_1 x^n + p_2 x^{n-1} + \ldots + p_n x + p_{n+1}$$

The self-pressurized experiments are fitted with a 5th order polynomial, whereas the actively pressurized experiments are fitted with a 4th order polynomial. The coefficients are listed in Table V-1, together with the fitting interval and the Root Mean Squared Error (RMSE). The coefficients are provided with high accuracy, because polynomial behaviour is very sensitive to these coefficients.

Exp	p_1	p_2	p_3	p_4	p_5	p_6	Fitting interval	RMSE [kPa]
sp120	4.337614484935347 e-006	-0.02206363000327	44.88290268596344	-4.564218740348201 e+004	2.320202334626482 e+007	-4.716605199305535 e+009	$t = 31$ s to $t = 100$ s	0.0368
sp130	1.365847128956317 e-006	-0.01018737591696	30.38483229098015	-4.529896495101380 e+004	3.375579587318180 e+007	-1.005801994266771 e+010	$t = 25$ s to $t = 100$ s	0.0258
sp140	3.529638195919456 e-006	-0.03915294429732	1.737103808210409 e+002	-3.853213523385930 e+005	4.273218174237453 e+008	-1.895440058956999 e+011	$t = 19.5$ s to $t = 100$ s	0.0303
sp160	1.861917040182717 e-006	-0.03536617388052	2.686986927015022 e+002	-1.020709751276620 e+006	1.938642250951756 e+009	-1.472791124088715 e+012	$t = 17$ s to $t = 100$ s	0.0277
ap120	3.654161113228058 e-004	-0.09250384822422	8.96613642455184	-5.089526680330116 e+002	1.321295165657648 e+005		$t = 10$ s to $t = 50$ s	0.041
ap130	0.00209888176451	-0.85638712491564	1.327494714232154 e+002	-9.440498149192687 e+003	3.844865685869895 e+005		$t = 10$ s to $t = 50$ s	0.033
ap140	0.00110409753383	-0.61318257272436	1.305343668491043 e+002	-1.280296058251615 e+004	6.168032091733239 e+005		$t = 10$ s to $t = 50$ s	0.028
ap150	0.00282996059216	-1.77554815039683	4.213696316919444 e+002	-4.510991607273714 e+004	1.980013321673860 e+006		$t = 5$ s to $t = 50$ s	0.042
ap160	0.00322553603822	-2.41573959355711	6.829652210512585 e+002	-8.674697042366408 e+004	4.331521628431309 e+006		$t = 5$ s to $t = 50$ s	0.041

Table V-1. Function data for fitting of the pressure curves

VI. Ullage temperature developments

This appendix gives the temperature data measured by each sensor in graphical form. Temperature at the four probe heights closest to the liquid surface is the average temperature of the two sensors at this height (for example the temperature at $z = 0.334$ m is calculated as $\dfrac{T_4 + T_8}{2}$).

Self pressurisation

Figure VI-1 to Figure VI-4 show the measured temperatures for each self-pressurized experiment. The temperature increases graduately during the self-pressurization phase. When sloshing starts the temperature at $z = 0.334$ m drops slightly. The temperatures in the other regions show only minor change.

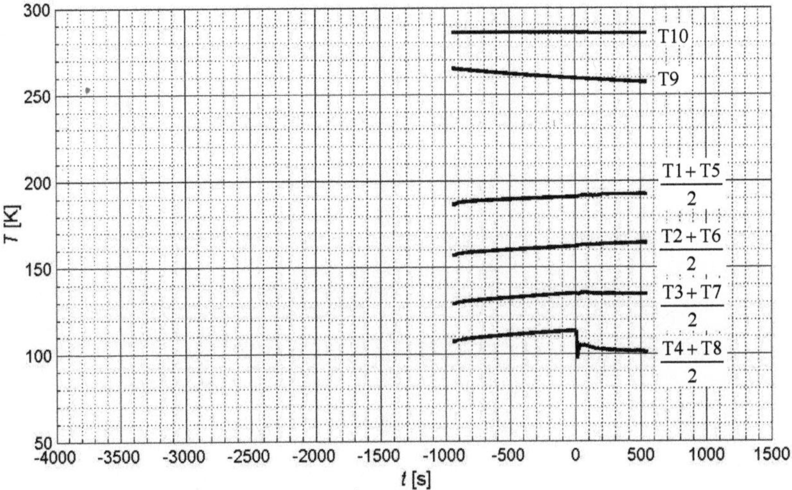

Figure VI-1. Ullage temperature developement for sp120.

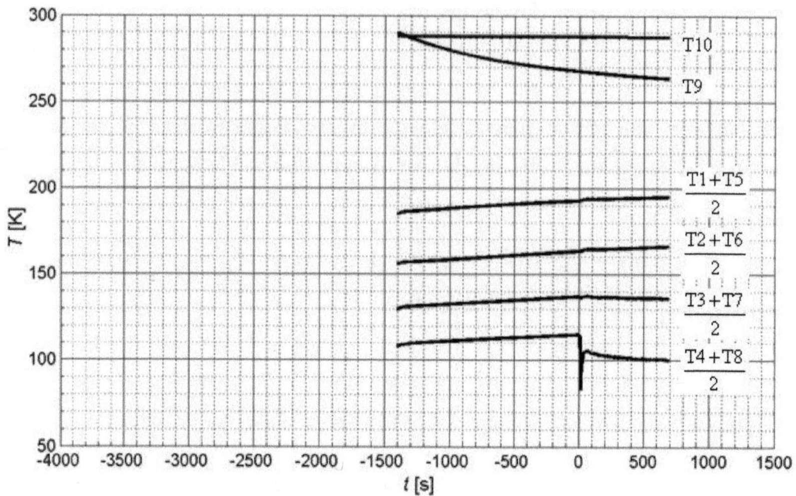

Figure VI-2. Ullage temperature developement for sp130.

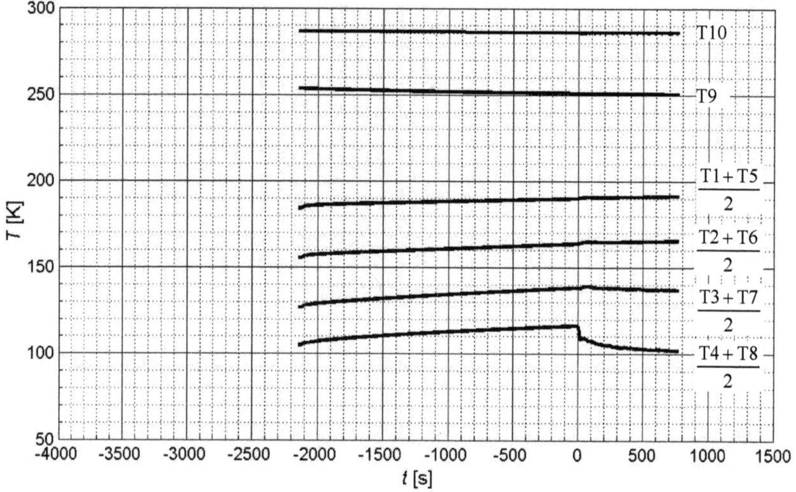

Figure VI-3. Ullage temperature developement for sp140.

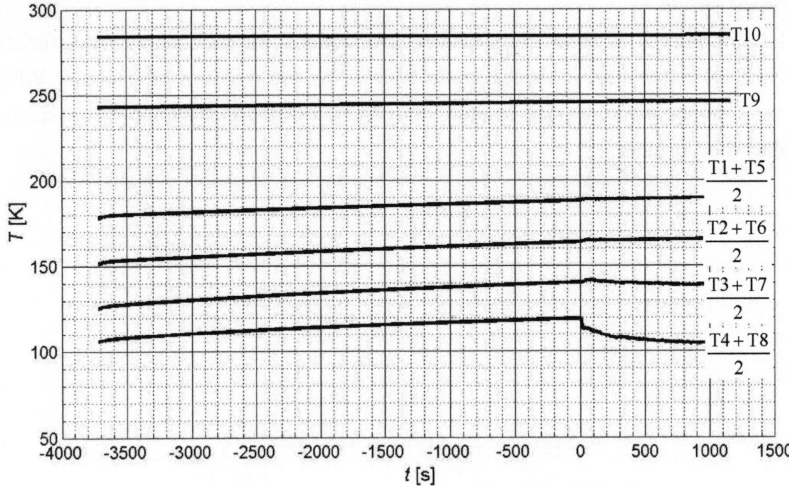

Figure VI-4. Ullage temperature developement for sp160.

Active pressurisation

Figure VI-1 to Figure VI-4 show the measured temperatures for each actively pressurized experiment. The temperature increases graduately during the pressurization phase. When sloshing starts the temperatures start to drop. Only the temperature at the lid (T10) remains unaffected by the sloshing.

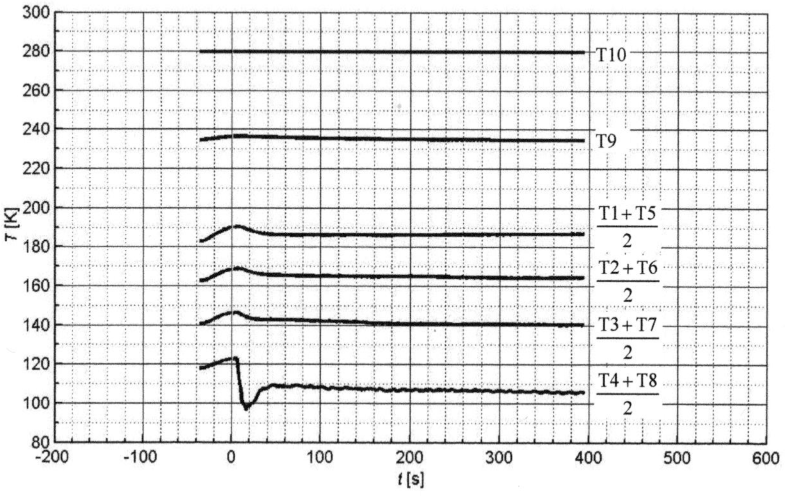

Figure VI-5. Ullage temperature developement for ap120.

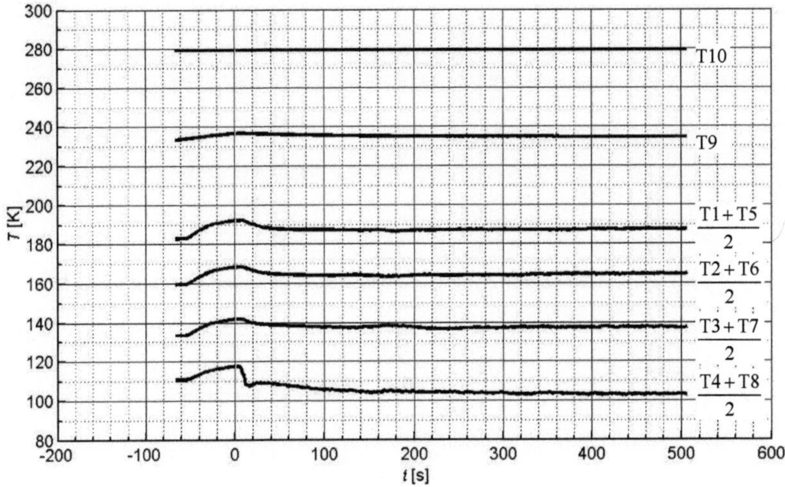

Figure VI-6. Ullage temperature developement for ap130.

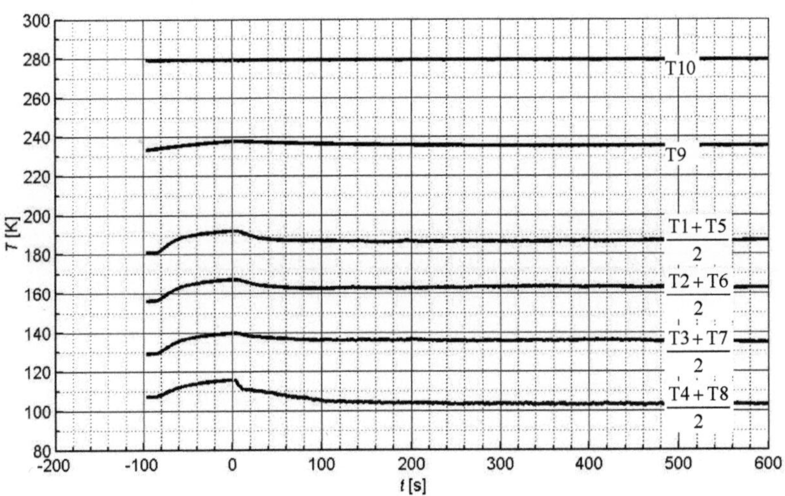

Figure VI-7. Ullage temperature developement for ap140.

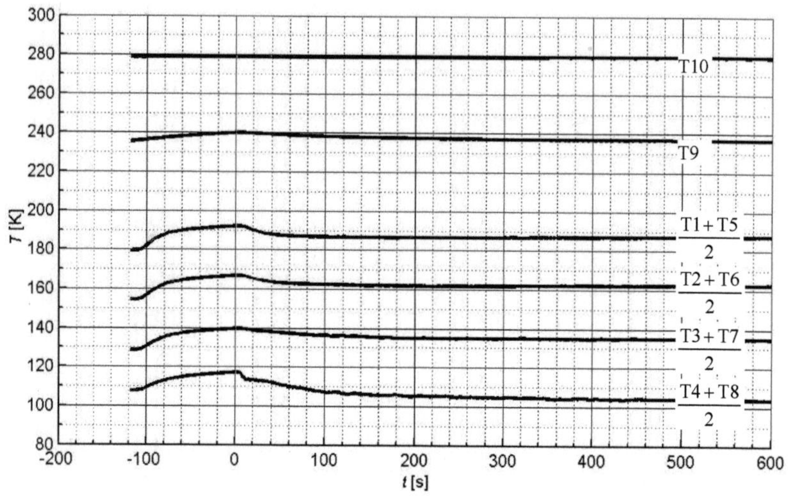

Figure VI-8. Ullage temperature developement for ap150.

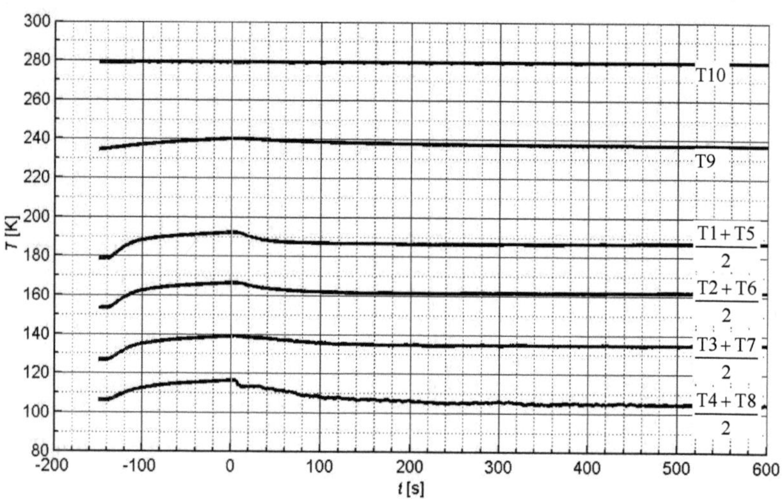

Figure VI-9. Ullage temperature developement for ap160.

VII.   Nitrogen property tables from NIST

Nitrogen at 100 kPa

T	p	ρ	i	h	c_v	c_p	β	μ	λ	$\widetilde{L}$
[K]	[kPa]	[kg m^{-3}]	[kJ kg^{-1}]	[kJ kg^{-1}]	[kJ K^{-1} kg^{-1}]	[kJ K^{-1} kg^{-1}]	[K^{-1}]	[µPa s^{-1}]	[W m^{-1} K^{-1}]	[kJ kg^{-1}]
76.5	100	810.0	-123.886	-123.762	1.08910	2.03742	0.0056	166.2	0.15	
76.6	100	809.5	-123.682	-123.559	1.08850	2.03788	0.0056	165.6	0.15	
76.7	100	809.1	-123.478	-123.355	1.08791	2.03835	0.0056	164.9	0.15	
76.8	100	808.6	-123.275	-123.151	1.08732	2.03882	0.0056	164.3	0.15	
76.9	100	808.2	-123.071	-122.947	1.08673	2.03929	0.0056	163.6	0.15	
77.0	100	807.7	-122.867	-122.743	1.08614	2.03977	0.0056	162.9	0.15	
77.1	100	807.2	-122.663	-122.539	1.08555	2.04026	0.0056	162.3	0.15	
77.2	100	806.8	-122.459	-122.335	1.08497	2.04074	0.0057	161.6	0.15	
77.24	100	806.6	-122.370	-122.246	1.08471	2.04095	0.0057	161.4	0.15	199.3
77.24	100	4.6	55.1	77.1	0.77	1.12	0.0147	5.4	0.01	199.3
78	100	4.5	55.7	77.9	0.77	1.12	0.0145	5.5	0.01	
79	100	4.4	56.5	79.0	0.77	1.12	0.0143	5.5	0.01	
80	100	4.4	57.3	80.2	0.77	1.11	0.0140	5.6	0.01	
90	100	3.8	65.1	91.1	0.76	1.09	0.0120	6.3	0.01	
100	100	3.4	72.8	101.9	0.75	1.07	0.0106	7.0	0.01	
110	100	3.1	80.4	112.6	0.75	1.06	0.0095	7.6	0.01	
120	100	2.8	88.0	123.2	0.75	1.06	0.0086	8.3	0.01	
130	100	2.6	95.5	133.7	0.75	1.05	0.0079	8.9	0.01	
140	100	2.4	103.0	144.2	0.74	1.05	0.0073	9.5	0.01	
150	100	2.3	110.5	154.7	0.74	1.05	0.0068	10.1	0.01	
160	100	2.1	117.9	165.2	0.74	1.05	0.0063	10.7	0.02	
170	100	2.0	125.4	175.7	0.74	1.05	0.0060	11.3	0.02	
180	100	1.9	132.9	186.1	0.74	1.04	0.0056	11.8	0.02	
190	100	1.8	140.3	196.6	0.74	1.04	0.0053	12.4	0.02	
200	100	1.7	147.8	207.0	0.74	1.04	0.0050	12.9	0.02	
210	100	1.6	155.2	217.4	0.74	1.04	0.0048	13.5	0.02	
220	100	1.5	162.7	227.9	0.74	1.04	0.0046	14.0	0.02	
230	100	1.5	170.1	238.3	0.74	1.04	0.0044	14.5	0.02	
240	100	1.4	177.5	248.7	0.74	1.04	0.0042	15.0	0.02	
250	100	1.3	185.0	259.1	0.74	1.04	0.0040	15.5	0.02	
260	100	1.3	192.4	269.5	0.74	1.04	0.0039	16.0	0.02	
270	100	1.2	199.9	280.0	0.74	1.04	0.0037	16.5	0.02	
280	100	1.2	207.3	290.4	0.74	1.04	0.0036	17.0	0.02	
290	100	1.2	214.7	300.8	0.74	1.04	0.0035	17.4	0.03	
300	100	1.1	222.2	311.2	0.74	1.04	0.0033	17.9	0.03	

Nitrogen at 160 kPa

T [K]	p [kPa]	ρ [kg m^{-3}]	i [kJ kg^{-1}]	h [kJ kg^{-1}]	c_v [kJ K^{-1} kg^{-1}]	c_p [kJ K^{-1} kg^{-1}]	β [K^{-1}]	μ [μPa s^{-1}]	λ [W m^{-1} K^{-1}]	$\widetilde{L}$ [kJ kg^{-1}]
77	160	807.9	-122.900	-122.702	1.08630	203911	0.0056	163.1	0.15	
77.1	160	807.4	-122.696	-122.498	1.08571	2.03959	0.0056	162.5	0.15	
77.2	160	806.9	-122.492	-122.294	1.08513	2.04007	0.0057	161.8	0.15	
77.3	160	806.5	-122.288	-122.090	1.08454	2.04056	0.0057	161.2	0.15	
77.4	160	806.0	-122.084	-121.886	1.08396	2.04105	0.0057	160.5	0.15	
77.5	160	805.6	-121.880	-121.681	1.08337	2.04155	0.0057	159.9	0.15	
77.6	160	805.1	-121.676	-121.477	1.08279	2.04204	0.0057	159.3	0.15	
77.7	160	804.7	-121.472	-121.273	1.08221	2.04254	0.0057	158.6	0.15	
77.8	160	804.2	-121.268	-121.069	1.08163	2.04305	0.0057	158.0	0.14	
77.9	160	803.7	-121.063	-120.864	1.08105	2.04356	0.0057	157.4	0.14	
78	160	803.3	-120.859	-120.660	1.08047	2.04407	0.0057	156.8	0.14	
78.1	160	802.8	-120.655	-120.456	1.07989	2.04459	0.0057	156.1	0.14	
78.2	160	802.4	-120.450	-120.251	1.07932	2.04511	0.0057	155.5	0.14	
78.3	160	801.9	-120.246	-120.047	1.07874	2.04564	0.0057	154.9	0.14	
78.4	160	801.4	-120.042	-119.842	1.07817	2.04617	0.0058	154.3	0.14	
78.5	160	801.0	-119.837	-119.637	1.07760	2.04670	0.0058	153.7	0.14	
78.6	160	800.5	-119.632	-119.433	1.07702	2.04724	0.0058	153.1	0.14	
78.7	160	800.1	-119.428	-119.228	1.07645	2.04778	0.0058	152.5	0.14	
78.8	160	799.6	-119.223	-119.023	1.07588	2.04833	0.0058	151.9	0.14	
78.9	160	799.1	-119.018	-118.818	1.07532	2.04888	0.0058	151.4	0.14	
79	160	798.7	-118.814	-118.613	1.07475	2.04943	0.0058	150.8	0.14	
79.1	160	798.2	-118.609	-118.408	1.07418	2.04999	0.0058	150.2	0.14	
79.2	160	797.7	-118.404	-118.203	1.07362	2.05056	0.0058	149.6	0.14	
79.3	160	797.3	-118.199	-117.998	1.07305	2.05113	0.0058	149.0	0.14	
79.4	160	796.8	-117.994	-117.793	1.07249	2.05170	0.0058	148.5	0.14	
79.5	160	796.3	-117.789	-117.588	1.07193	2.05227	0.0059	147.9	0.14	
79.6	160	795.9	-117.584	-117.383	1.07137	2.05286	0.0059	147.3	0.14	
79.7	160	795.4	-117.378	-117.177	1.07081	2.05344	0.0059	146.8	0.14	
79.8	160	794.9	-117.173	-116.972	1.07025	2.05403	0.0059	146.2	0.14	
79.9	160	794.5	-116.968	-116.766	1.06969	2.05463	0.0059	145.7	0.14	
80	160	794.0	-116.762	-116.561	1.06913	2.05523	0.0059	145.1	0.14	
80.1	160	793.5	-116.557	-116.355	1.06858	2.05583	0.0059	144.6	0.14	
80.2	160	793.1	-116.352	-116.150	1.06802	2.05644	0.0059	144.0	0.14	
80.3	160	792.6	-116.146	-115.944	1.06747	2.05705	0.0059	143.5	0.14	
80.4	160	792.1	-115.940	-115.738	1.06692	2.05767	0.0059	142.9	0.14	
80.5	160	791.7	-115.735	-115.533	1.06637	2.05829	0.0060	142.4	0.14	
80.6	160	791.2	-115.529	-115.327	1.06582	2.05892	0.0060	141.9	0.14	
80.7	160	790.7	-115.323	-115.121	1.06527	2.05955	0.0060	141.3	0.14	
80.8	160	790.2	-115.117	-114.915	1.06472	2.06019	0.0060	140.8	0.14	
80.9	160	789.8	-114.911	-114.709	1.06417	2.06083	0.0060	140.3	0.14	
81	160	789.3	-114.705	-114.503	1.06363	2.06148	0.0060	139.8	0.14	

81.1	160	788.8	-114.499	-114.296	1.06308	2.06213	0.0060	139.2	0.14		
81.2	160	788.3	-114.293	-114.090	1.06254	2.06279	0.0060	138.7	0.14		
81.3	160	787.9	-114.087	-113.884	1.06200	2.06345	0.0060	138.2	0.14		
81.4	160	787.4	-113.881	-113.678	1.06146	2.06412	0.0060	137.7	0.14		
81.45	160	787.1	-113.775	-113.572	1.06118	2.06446	0.0060	137.4	0.14	193.7	
81.45	160	7.0	57.4	80.1	0.78	1.16	0.0147	5.8	0.01	193.7	
82	160	7.0	57.8	80.7	0.78	1.15	0.0146	5.8	0.01		
83	160	6.9	58.6	81.9	0.78	1.15	0.0143	5.9	0.01		
84	160	6.8	59.5	83.0	0.78	1.14	0.0140	5.9	0.01		
85	160	6.7	60.3	84.2	0.77	1.14	0.0138	6.0	0.01		
86	160	6.6	61.1	85.3	0.77	1.13	0.0135	6.1	0.01		
87	160	6.5	61.9	86.4	0.77	1.13	0.0133	6.1	0.01		
88	160	6.4	62.7	87.6	0.77	1.13	0.0131	6.2	0.01		
89	160	6.3	63.5	88.7	0.77	1.12	0.0129	6.3	0.01		
90	160	6.3	64.3	89.8	0.77	1.12	0.0127	6.3	0.01		
100	160	5.6	72.1	100.9	0.76	1.09	0.0110	7.0	0.01		
110	160	5.0	79.8	111.7	0.75	1.08	0.0098	7.6	0.01		
120	160	4.6	87.5	122.4	0.75	1.07	0.0088	8.3	0.01		
130	160	4.2	95.1	133.1	0.75	1.06	0.0080	8.9	0.01		
140	160	3.9	102.6	143.7	0.75	1.06	0.0074	9.5	0.01		
150	160	3.6	110.1	154.2	0.75	1.05	0.0069	10.1	0.01		
160	160	3.4	117.6	164.8	0.74	1.05	0.0064	10.7	0.02		
170	160	3.2	125.1	175.3	0.74	1.05	0.0060	11.3	0.02		
180	160	3.0	132.6	185.8	0.74	1.05	0.0057	11.8	0.02		
190	160	2.8	140.1	196.2	0.74	1.05	0.0053	12.4	0.02		
200	160	2.7	147.6	206.7	0.74	1.05	0.0051	12.9	0.02		
210	160	2.6	155.0	217.2	0.74	1.05	0.0048	13.5	0.02		
220	160	2.5	162.5	227.6	0.74	1.04	0.0046	14.0	0.02		
230	160	2.3	169.9	238.1	0.74	1.04	0.0044	14.5	0.02		
240	160	2.2	177.4	248.5	0.74	1.04	0.0042	15.0	0.02		
250	160	2.2	184.8	258.9	0.74	1.04	0.0040	15.5	0.02		
260	160	2.1	192.3	269.4	0.74	1.04	0.0039	16.0	0.02		
270	160	2.0	199.7	279.8	0.74	1.04	0.0037	16.5	0.02		
280	160	1.9	207.2	290.2	0.74	1.04	0.0036	17.0	0.02		
290	160	1.9	214.6	300.6	0.74	1.04	0.0035	17.4	0.03		
300	160	1.8	222.0	311.1	0.74	1.04	0.0033	17.9	0.03		

VIII.   Borosilicate material properties from [40]

The density can be assumed temperature independent at $\rho = 2230$ kg m^{-3}.

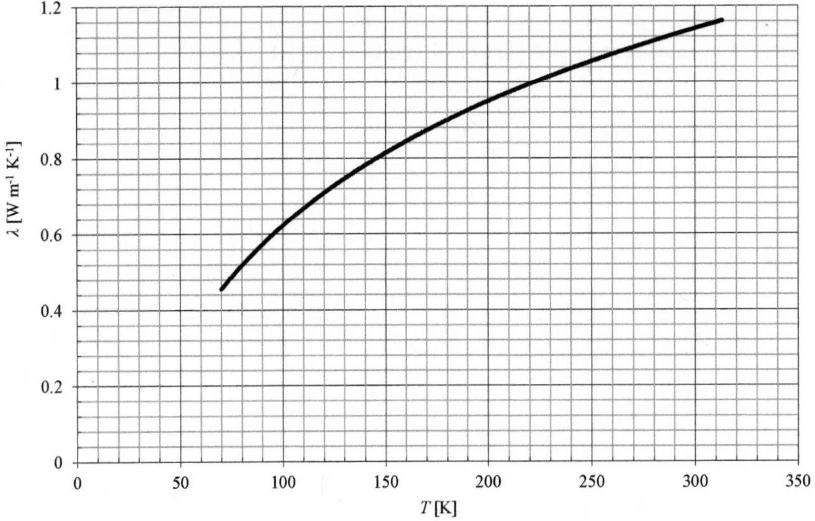

Figure VIII-1. Borosilicate heat conductivity.

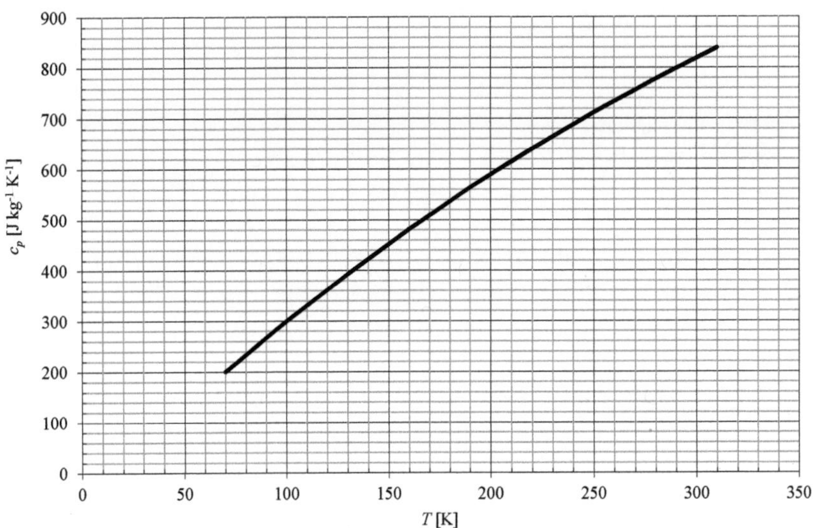

Figure VIII-2. Specific heat capacity of borosilicate.

	λ [Wm^{-1}K^{-1}]	ρ [kg m^{-3}]	c [J kg^{-1} K^{-1}]
Silicone @ 293 K	0.2	1500	1500
Polyacetal @ 293 K	0.3	1400	1500

Table VIII-1. Lid (polyacetal) and gasket (silicone) material properties.